The Martians

ALSO BY DAVID BARON

American Eclipse

The Beast in the Garden

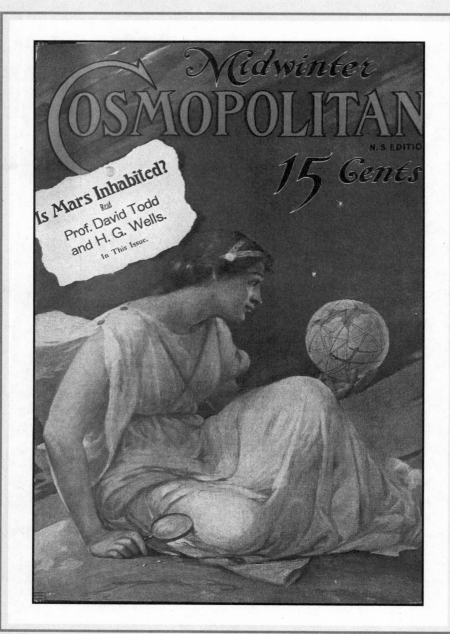

THE MARTIANS

THE TRUE STORY OF AN ALIEN CRAZE THAT CAPTURED TURN-OF-THE-CENTURY AMERICA

DAVID BARON

Liveright Publishing Corperation
A Division of W. W. Norton & Company
Independent Publishers Since 1923

Copyright © 2025 by David Baron

All rights reserved
Printed in the United States of America
First Edition

For information about permission to reproduce selections from this book, write to
Permissions, Liveright Publishing Corporation, a division of W. W. Norton
& Company, Inc., 500 Fifth Avenue, New York, NY 10110

For information about special discounts for bulk purchases, please contact
W. W. Norton Special Sales at specialsales@wwnorton.com or 800-233-4830

Manufacturing by Lakeside Book Company
Book design by Barbara Bachman
Production manager: Lauren Abbate

ISBN 978-1-324-09066-3

Liveright Publishing Corporation
500 Fifth Avenue, New York, NY 10110
www.wwnorton.com

W. W. Norton & Company Ltd.
15 Carlisle Street, London W1D 3BS

10 9 8 7 6 5 4 3 2 1

**To my
nieces and nephews.**

*May your generation be
the first to cross the ocean of
interplanetary space.*

Mankind has to all intents and purposes been journeying Marsward through the years.

—PERCIVAL LOWELL

Contents

PROLOGUE: Why Mars? ... *1*

· Part One ·
CENTURY'S END · 1876–1900

CHAPTER 1	Evolution	*9*
CHAPTER 2	The French Philosopher	*22*
CHAPTER 3	What the Colorblind Astronomer Saw	*34*
CHAPTER 4	Mars Hill	*46*
CHAPTER 5	Lost World	*58*
CHAPTER 6	Allies and Adversaries	*69*
CHAPTER 7	War Stories	*79*
CHAPTER 8	Wireless	*90*

· Part Two ·
A NEW CIVILIZATION · 1901–1907

CHAPTER 9	Messengers	*105*
CHAPTER 10	"Small Boy Theory"	*119*
CHAPTER 11	The Martians of Earth	*130*
CHAPTER 12	Truth in the Negative	*141*
CHAPTER 13	Planet of Peace	*152*
CHAPTER 14	Alianza	*162*

· Part Three ·
THE EARTHLINGS RESPOND · 1908–1916

CHAPTER 15	Articles of Faith	*183*
CHAPTER 16	Skyward	*192*
CHAPTER 17	Endgame	*204*
CHAPTER 18	Poetic Achievement	*222*
	EPILOGUE: Children of Mars	*232*

ACKNOWLEDGMENTS	*243*
NOTES ON SOURCES	*249*
SELECT BIBLIOGRAPHY	*289*
LIST OF ILLUSTRATIONS	*305*
INDEX	*311*

The Martians

Prologue

Why Mars?

By the time the Mars craze reached its zenith, toward the end of the twentieth century's first decade, one could scarcely avoid the Martians even in the stately parlors and ballrooms of New York's rarefied society.

Mamie Fish, a legendary hostess famed for her acid wit and outlandish functions, threw a dinner for twenty in the winter of 1907. Between courses, a dweller from the red planet materialized through scarlet drapes and boomed to the bewildered guests, "Mortals, I am here to tell you the things we know of you in Mars." Across town, at the J. Hamilton P. Hodgsons, another Martian dropped by for New Year's Eve. Though unable to speak English, he was said to converse with partygoers by means of a so-called mystiscope, a device that helpfully enabled translation. A year or two earlier, at the enormous Midtown playhouse known as the New York Hippodrome, some of the handsomest couples in the city—husbands in evening dress, wives in flowing gowns and pearls—had filled the boxes for the premiere of *A Yankee Circus on Mars*. On a stage that stretched to seemingly astronomical proportions, acrobats and bareback riders performed amid a chorus of Martians three hundred strong.

The Martians that appeared in Manhattan's theaters and elite drawing rooms were—one might reasonably assume—mere humans in alien attire, but those that showed up in the magazines and daily papers surely were not. Artists' renditions depicted the Martians the way scientists imagined their environment had molded them: tall and lithe, perhaps winged or gilled, sporting an insect's antennae or an elephant's trunk. The illustrations left readers agape and aghast. "The Martians may be very fine intellectual people," remarked one Jane Jones of Binghamton, New York, "but we do not, and never shall like their looks."

Although their visages were homely, the physical infrastructure they had erected was something to behold. Telescopic observations revealed on Mars "great centres of population" and "a network of marvelously designed canals traversing the deserts," *The New York Times* reported in a story that was hardly meant to be a hoax. Photographs corroborated the existence of such monumental feats of hydraulic engineering. The New York *World* declared in a headline, SCIENTISTS NOW KNOW POSITIVELY THAT THERE ARE THIRSTY PEOPLE ON MARS, and among those who knew this positively was Alexander Graham Bell, America's elder statesman of science. Having reviewed the evidence, he saw "no escape from the conviction that Mars is inhabited by a highly civilized and intelligent race of beings."

Meanwhile, towering above the trees on Long Island's north shore stood the instrument that might soon enable humankind to contact the Martians, an interplanetary radio transmitter that had been financed by J. P. Morgan, the Wall Street titan and one of the nation's most sober moneymen. Philosophers, spiritualists, even God-fearing clergy craved to learn what those sapient aliens across the void might teach us. "*Some day theology will be shaken out of its seed-pod present of earth-centrocism and egotism by the study of soul-life on this brother-planet,*" a Chicago minister wrote with avid expectation to the astronomer who, more than anyone, had staked his career on understanding the Martians and their society. "So toil away, and God bless you!"

For a decade and a half, the Mars craze spread like a raging epidemic—or a burgeoning wildfire—across class divides and borders, but particularly in the United States, consuming almost everything in its path. In the midst of it all, as if in the fog of a long, drawn-out battle, it was often difficult to discern the truth, to separate insight from delusion, to identify who were the geniuses and who were the cranks. By the time the frenzy abated in the years around World War I, some of the glorified participants would end up publicly humiliated and bankrupt, even institutionalized, and the world's perception of its place in the universe would be forever altered.

...

I ARRIVED ON EARTH in the 1960s, and, like many children born in the age of NASA's Apollo program, I was raised on spacemen and Martians. My first toy action figure was a crew-cut astronaut with bendable limbs and an American flag emblazoned on his chest, and I can still picture that electrifying summer night in 1969 when my family gathered to watch Neil Armstrong step onto the lunar surface. It was on TV that I saw Martians, too, though I knew—or had been told anyway—that they were pure make-believe. I laughed at Ray Walston in *My Favorite Martian*, a sitcom about a wayward alien who had crash-landed in California and who sprouted antennae from his head, for reasons I failed to comprehend yet readily accepted. Watching Saturday-morning cartoons, I rooted for Bugs Bunny as he battled Marvin the Martian's hapless attempts at destroying Earth. I also encountered hand-me-down Martians, ones from my parents' generation that I discovered in old comic books and movies and sci-fi novels.

More than a half century later, having built a career as a science writer specializing in astronomy, I have not lost my passion for Mars or for space exploration. Neither has my country. Today, NASA scientists and private entrepreneurs are actively planning to send astronauts to the red planet, and our robots are already on the Martian surface—analyzing the soil, probing for water, sniffing the air. Radical visionaries aim to populate Mars with domed cities provisioned by farmers in inflatable greenhouses, creating a new world that could serve as civilization's escape hatch should a cosmic cataclysm make our home planet uninhabitable. For as long as I—indeed, as long as any of us living—can remember, Mars has loomed large in the public consciousness. This seems odd for such a diminutive world in a solar system filled with planets and moons, asteroids and comets. There is something about Mars that transcends all other astronomical bodies;

it possesses an undeniable aura of mystery and romance, an allure not fully explained by its physical reality.

Why Mars? Where did our infatuation come from?

As I discovered when writing this book, much of it began in a strange era, not that long ago, when it was the astronauts who were fiction and the Martians that were real.

...

BEFORE I RAISE THE curtain, a few preliminaries.

This is a drama of planetary scale and operatic scope, a story as much about Earth and earthlings as it is about Mars and Martians, for the two worlds—and their two peoples—tightly intertwined during the era I chronicle. The expansive cast includes a clutch of noted astronomers: an Italian whose maps of Mars first puzzled, then galvanized, the world; a Frenchman whose fertile mind infused the red planet with romance; an American for whom Mars became a Rorschach test, revealing more about his inner self than outer space. Others were writers: a British novelist whose visions of Mars induced nightmares on Earth; a cultured Bostonian who lent her class and credibility to the Martian enterprise; a Brooklyn-based journalist who became a trusted guide, helping the public interpret the astonishing discoveries on Mars. Joining this crew was a genius inventor obsessed with contacting the Martians by wireless, and an adventurous couple—husband and wife—who trekked five thousand miles with a seven-ton telescope to secure the stunning evidence that would propel the Mars craze to its fevered climax.

I, too, will appear in these pages, for the writing of this book was itself a journey of exploration, one that crossed three continents as I aimed to uncover the tale's plot, familiarize myself with its characters, deduce its lessons. Nothing in the chapters that follow has been invented or embellished. All details and quotations have been culled from a review of the historical record comprising hundreds of books, thousands of newspaper and magazine articles, and tens of thousands of pages in diaries, scrapbooks, observation logs, and correspondence held in myriad archives. Readers familiar with the history of Mars will find many new details here, including previously unrecognized points of intersection among the protagonists and novel connections between what was happening on Earth and what the public imagined on Mars.

One may interpret this narrative in multiple ways. It is an inspiring epic of human inventiveness. It is a cautionary tale of mass delusion. It is a drama of battling egos. Ultimately, though, it is a love story, an account of when we, the people of Earth, fell hard for another planet and projected our fantasies, desires, and ambitions onto an alien world. This romance blazed before it turned to embers, and it produced children, for we—the first humans who may actually sail to Mars—are its descendants.

PART ONE

CENTURY'S END

1876–1900

CHAPTER

I

Evolution

1876–1891

Had mortal eyes peered down from space on that sweltering day in 1876, had they fixed their alien gaze toward Boston, they surely would have missed a cluster of humans who scurried about like mites, their bodies too small to be discerned through a telescope at interplanetary distance. Upon closer inspection, however, the Martians might well have inferred the presence of intelligent minds. The Earth's terrain revealed strange geometries. Just north of the Charles River's natural meanders, an array of straight lines—a road network—converged like spokes on a wheel. Near their common intersection, yet more narrow streaks—a lattice of footpaths—crisscrossed a leafy quadrangle. These shapes, to those watching from above, would have displayed evidence of a technological civilization. Yet to understand that civilization, to decipher its character, would have required more intimate scrutiny, would have necessitated the sending of an emissary.

It was a Wednesday in late June, and the morning saw a stream of carriages pull up to the gates of Harvard College. Women in bright dresses and men clad in more somber tones mingled on the grass with faculty in Oxford gowns. As the sun shone and a breeze blew and the crowd sought shade beneath the elms, the governor and his entourage paraded up the avenue in colorful splendor. The ceremonial guard—cavalrymen in red tunics—halted outside Massachusetts Hall, a brick

HARVARD UNIVERSITY, CAMBRIDGE, MASSACHUSETTS.

dormitory where John Hancock and the Adamses (Samuel and John) had spent their undergraduate days a century before. At this spot the academic procession assembled, and the long column, with degree candidates positioned toward the front, began its slow march across campus. Students and dignitaries filtered into the theater, where parents waited. A brass band and male chorus joined in stirring harmony as a solemn celebration began.

For the 235th time in its storied history, Harvard University was holding graduation exercises. Here gathered were the families of old New England: the Bradfords and Cabots, the Longfellows and Sargents, the Talbots and especially the Lowells. James Russell Lowell, class of 1838 and founding editor of the influential *Atlantic Monthly*, headed the Harvard Alumni Association. His cousin Augustus Lowell, class of '50 and a major benefactor of the university, was the association's treasurer. And Augustus's firstborn son, a graduating senior, had been asked to deliver a disquisition—an essay—at commencement.

Tall and blue-eyed, Percival Lowell took the shallow stage and began to recite. "As a traveler wandering through a forest may read there its whole history, from the time when each tree was a young and tender sapling until it grew to be an old and knotty trunk; so may an astronomer in his nightly watchings of the heavens read there the past and future of every star." The audience sat arranged in a semicircle, perched on hard benches of black walnut. The women fanned their faces while this exceptionally poised young man gave a lecture that proved (as one in the crowd noted) "quite profound, but somewhat dry for such warm weather."

PERCIVAL LOWELL, HARVARD CLASS OF 1876.

Lowell addressed a weighty subject: the formation of the solar system. He described the latest thinking about how the sun and planets came to be, not directly at the hand of God but out of a spinning disk of gas and dust that spiraled and eddied into a string of separate worlds. Scientists assumed that since all of the planets had formed from the same raw material, they would follow the same life trajectory from molten birth through tranquil middle age to eventual decay and death, each at its own pace. "We can look on, beyond the present, to the time when our Earth shall no longer be an inhabited world," Lowell said, striking a bleak note on a day that celebrated new beginnings. "We can see how, in their allotted time, each of the planets shall follow the example of our Earth, until finally the sun himself shall have ended his work forever."

As it is with the creation of worlds, so it is with the emergence of worldviews. A society's understanding of itself and of its place in the universe coalesces out of a diffuse collection of ideas. These concepts interact in complex ways as notions evolve and clash, and movements begin. At Harvard University on June 28, 1876, such a process was commencing. The vortex was beginning to spin.

...

FOR MANY PEOPLE IN Gilded Age America and late Victorian England, the times proved exhilarating and unsettling. The secrets of nature, so long cloaked in myth and mystery, were succumbing to science. The steam engine and telegraph, and soon the incandescent light, conquered distance and darkness. Geological discoveries revealed the unfathomable depth of time, and those in astronomy the infinitude of space. Meanwhile, a unified philosophical framework had taken hold of the public, providing a new way of seeing the universe and its workings. "Everybody nowadays talks about evolution," a British magazine asserted. "Like electricity, the cholera germ, [and] woman's rights . . . it is 'in the air.'"

Evolution encompassed far more than Charles Darwin's theory of natural selection, which had convulsed so much of the Christian world when published in 1859. Whereas the Bible explained creation as a miraculous event performed in six days, evolutionary doctrine proposed that the physical world around us—everything from stars and oceans to mushrooms to man—came about through mundane processes operating over millions of years. To the British philosopher Herbert Spencer, whose writings deeply molded the public mind, evolution implied *progress*, an inexorable march from simplicity toward complexity that could be seen as well in human affairs—in cultures and nations, languages and religions, societies and individuals.

In the years after his Harvard graduation, Percival Lowell underwent several phases of his own rapid intellectual evolution. This zigzag path would eventually point toward Mars, but first he was set on a different course, the destination seemingly preordained. Lowell was a young man of privilege, a "Boston Brahmin," as Oliver Wendell Holmes Sr. dubbed New England's elite caste, an American aristocracy marked by "a provokingly easy way of dressing, walking, talking, and nodding to people, as if they felt entirely at home, and would not be embarrassed in the least, if they met . . . the President of the United

States, face to face." Brahmin families were refined and educated, not to mention intertwined. Cousins married cousins. Wealth begat wealth.

Lowell, like so many of his social class, descended from money on both sides of the family. His father's great-uncle had sown the seeds of America's Industrial Revolution after stealing the secrets of British manufacturing and planting them in New England. His father's father then grew the family company into an empire, erecting an entire town of cotton mills to be known as Lowell, Massachusetts. Percival's maternal grandfather, Abbott Lawrence, also made a fortune in textiles, establishing the neighboring namesake mill city of Lawrence, Massachusetts, and the affluence of both bloodlines further increased under the shrewd financial management of Percival Lowell's father. As firstborn son, Percival himself would have been expected to continue in the family's line of work, yet he had failed to inherit the temperament of a businessman.

"Here things are rather dull," he complained to his sister several years after college, when he was working in his grandfather's stodgy office for want of anything else to do. "Not that I quite go in for murder as a necessary stimulus, but people might learn to be lively and gay without being bad." Lowell appeared to be suffering the existential ennui of a young man who had too much. He was floundering, searching for deeper meaning out of life. "I have . . . become decidedly misanthropic," he confided to a Harvard chum, "and, with the exception of a few friends, should not feel many pangs at migrating to another planet."

...

IN THE LATTER HALF of the nineteenth century, travel and trade connected the continents like never before, and as the world seemingly shrank, Western nations grew infatuated with exotic animals, plants, and cultures. Zoos, botanical gardens, and books on foreign lands found enthusiastic audiences in Europe and America, so it was no surprise that in Boston, in the 1880s, a lecture series on East Asia drew overflow crowds.

Much like Victorian urban dwellers in England, the residents of

THE MASSACHUSETTS INSTITUTE OF TECHNOLOGY, BOYLSTON STREET.

Boston long enjoyed attending educational courses, and none were more popular than those offered by the Lowell Institute. Endowed by one of Percival Lowell's late cousins, it held free talks in a stout neo-Renaissance building in the Back Bay neighborhood, the first structure in a school called the Massachusetts Institute of Technology, which had been founded at the time of the Civil War. For six weeks in the winter of 1882, on Monday and Saturday evenings, eager lecture-goers flocked to the building through snow and cold rain, climbed the broad steps to a skylit atrium, then ascended to a large hall on the second floor. They came to hear Edward S. Morse, a prominent zoologist with an ample beard and infectious manner, who had recently spent two years living in Japan. He had gone there to study marine invertebrates but developed a fascination with the nation's people. "Going to Japan is like visiting a new world—another planet," he was wont to say.

EDWARD S. MORSE.

Japan had long isolated itself from the West, frozen in a state of feudal antiquity, but it was now opening up to foreign trade and customs. "European dress and ideas have gained a foothold among the wealthy," Morse told a packed audience at MIT, yet he emphasized that in many ways "the Japan of today is the Japan of several centuries ago." Morse was ambidextrous and artistic, a demon at the blackboard who drew sketches with both hands simultaneously. Chalk flew as he illustrated his descriptions of Japanese manners, architecture, and modes of transportation. His lectures mesmerized and amused. "No other of the several winter courses has been so thronged," the *Boston Evening Transcript* reported, and among those in the crowd, spellbound, was Percival Lowell. There, in that lecture hall, the melancholic son of privilege found his calling and, soon, his escape after he suffered a psychological breakdown, one of several that would punctuate his life.

...

IT MAY BE DIFFICULT to empathize with the mental travails of a pampered aristocrat from the Gilded Age. Lowell was, after all, a young man who wanted for nothing, who could do anything. Yet, as I delved into his background, I came to appreciate the prison he inhabited. For the scion of a famous lineage, the expectations weighed mightily. Lowell's stern father taught his sons that they must work hard in life and commit themselves to something "of real significance." Lowell lived in a world constrained by duty, manners, and blood.

What exists of Lowell's personal correspondence today resides in several archives, including collections of the Massachusetts Historical Society, which occupies an elegant limestone-and-yellow-brick structure on the edge of Boston's Back Bay. I had visited early in my research and thought I had reviewed everything of relevance, but when I learned later of an as-yet hidden tranche of the Lowell family's papers—unprocessed ones, never before seen by scholars—I readily returned. After arriving at the society's library and stowing my belongings in a locker (including ink pens, which might damage irreplaceable manuscripts), I stepped into the hushed reading room. The space itself felt as if from another time, with oversized oil portraits of the dead watching from walls of dark wood. A librarian emerged from the back with the archival materials I had requested, manila folders containing handwritten letters. I carefully lifted each page and found myself lost in a breathless family drama of injured pride and public obloquy.

For several years after his graduation from Harvard, Percival Lowell had acceded to social dictates and courted a suitably Brahmin young woman. Rose Lee was the daughter of a Boston investment banker and the sister-in-law of a rising New York politician named Theodore Roosevelt, and by the summer of 1882, Lowell took the next logical step when he proposed marriage. She accepted, and he announced the joyous news, but over the subsequent two weeks, Lowell grew disconsolate and withdrawn. At last, he concluded that he did not love Rose Lee as he should a future wife. Shockingly, he retracted the marriage proposal.

"Sir," read a note that arrived from Theodore Roosevelt. "It is with great pleasure I inform you that your conduct has been such as only a mean-spirited and cowardly blackguard could be guilty of." More derision came from closer to home. Frank Peabody, who was both a cousin-in-law of Lowell's and a cousin of Rose Lee, wrote disparagingly to Lowell's father, "[Percival] is responsible for having won a woman's affections without being certain of his own, and according to every rule of justice and honor should also be held responsible for the consequences." Those consequences included a public shaming, which Peabody heaped on Lowell for months, *cutting* him—that is, refusing to acknowledge his presence in society. The dispute metastasized into a monumental family quarrel as cousins, aunts, and uncles lined up on either side of the rift, charging one another with deceit, unmanliness, connivance, and slander.

In a letter to his parents, Percival Lowell recognized the mess he had caused. "Oh mother forgive me. I did not mean to do so much wrong," he pleaded. "Your sad, erring boy."

One can imagine the repercussions. Subject to gossip and insults, Lowell must have struggled to navigate his usual social circles. He had taken one unthinkable action—revoking his proposal—and now he resolved to take another. In the spring of 1883, he stepped away from the family business and, inspired by Edward Morse's lecture series, sailed for Japan to reinvent himself.

...

A MERE FOUR MONTHS after Lowell's departure for Asia, a diplomatic mission from the Far East arrived in the United States. This mysterious band of dignitaries, like visitors from outer space, came from one of the most reclusive nations on Earth and soon turned up at one of the most exclusive addresses in New York City: the Fifth Avenue Hotel. The marble edifice was known for its rich upholstery, gilded wood, and Republican powerbrokers who plied their influence in its corridors.

At eleven o'clock on a sunny morning in mid-September, excitement coursed through the building as hotel employees blocked off a main hallway. "Women, children, and men crowded behind the porters, and stared at the long expanse of carpet," noted one of many journalists who had come to observe. Suddenly, at the staircase, appeared an assemblage of men draped elegantly in silk. They wore flowing robes with hanging sleeves of different hues, so that when they walked it was "like watching a kaleidoscope." This was Korea's first diplomatic mission to America, a major step in the meeting of East and West.

In the broad hall of the hotel, in view of the crush, the Koreans approached a parlor on the south side of the building. They sank to their knees, fell prostrate to the carpet, then arose and advanced. Chester A. Arthur, twenty-first president of the United

PRESIDENT ARTHUR RECEIVING THE COREAN EMBASSY AT THE FIFTH AVENUE HOTEL.

States—who held court here when in town—extended a hand to the visitors. The head of the delegation, Prince Min Yong Ik, offered formal greetings in Korean while beside him, preparing to read the translation, stood a "ruddy-faced young American." It was Percival Lowell, face-to-face with America's president in appropriate Brahmin style.

Lowell was there as a formal part of the Korean delegation. He served as its "foreign secretary," a job that materialized when the officials from Seoul, on the way to the United States, stopped in Japan and sought an American guide. ("I am for the nonce a Corean," Lowell wrote to an old Harvard friend while on the ship to America, spelling "Korea" with a *c*, as was customary at the time.) The appointment of the young Bostonian struck many as puzzling, since he had never been to Korea. In fact, almost no Westerners had visited that part of Asia—and for good reason, Lowell told the *New-York Tribune*: "There has until recently been a death penalty enforced against any foreigner entering the land." The shroud of secrecy was starting to lift, however. The kingdom had now negotiated a treaty with the United States, prompting this visit.

For seven weeks, Lowell took the Korean diplomats on an extended

PERCIVAL LOWELL,   HONG YONG SIK,   MIN YONG IK,   SO KOANG POM,
FOREIGN SECRETARY.   VICE-MINISTER.   MINISTER.   SECRETARY OF EMBASSY.

THE COREAN EMBASSY TO THE UNITED STATES.

U.S. tour: to the White House, West Point, the textile mills of Lowell and Lawrence, Massachusetts. Throughout the journey, the young Bostonian helped the foreigners navigate a strange land and served as their spokesman, interpreting for Americans the customs and intentions of the otherworldly visitors. No longer a jaded businessman, Lowell was now an enthusiastic cultural liaison. The transformation had occurred so suddenly that no one back home had been prepared for his arrival as an honorary Korean. "His parents knew nothing," the matriarch of a noted Brahmin family gossiped in a letter to her son. "So closes Chap 2^d of the adventures of Percival Lowell."

...

HIS NEXT BOLD CHAPTERS came in quick succession.

Sailing back to Asia with the Korean diplomats, Lowell arrived at the gates of Seoul as guest of the king. A crenellated wall girdled the city, running up and down the surrounding hills. The dirt streets seemed to surge with people dressed "as imagination might paint the denizens of another planet," Lowell wrote with the affected prose typical of explorers of the era. He spent three months in Korea, one of the first Americans permitted a close inspection of the exotic land. One evening, at dusk, he gazed toward a mountain south of Seoul and watched as tiny lights appeared on the summit, poised "so high in the heavens, they might well be the light from other worlds." These were signal fires, used to transmit messages from peak to peak across the nation, a kind of visual telegraphy. On this night, the lights announced a comforting "all is well" through the silence of the dark.

Lowell the world traveler soon became Lowell the published author. After returning to Boston—where he re-entered polite society, the ruckus over his canceled engagement having died down—he wrote a travelogue of Korea. It was a lyrical if not quintessentially Eurocentric book about the strange things he had seen, which offered Western readers an opportunity for reflection—quite literally, because most everything in the country seemed as if in a mirror. "For any people to write backwards, to talk backwards, to sit upon their feet, to take off not their hats but their shoes on entering a house, and in countless other ways to conform to what seems more like a photographic negative of our own civilization than a companion picture," Lowell wrote, "this is indeed a social phenomenon to rouse the most sluggish curiosity." More than curiosity, the book roused praise. *The Atlantic Monthly* commended Lowell's sympathetic treatment of "an alien race," as it termed the Korean people.

Now in his thirties, with his youthful mistakes behind him, Lowell matured into a man of letters and a man-about-town. He composed essays and poems, and wrote more books, all the while appearing in the society pages for his sportsmanship at polo and attendance at Gilded Age parties and weddings. (He served as best man for the marriage of Edith Wharton, whose novels came to reflect the rites and vanities of the cultural elite.) Lowell grew self-assured and could be arrogant, but he retained a winning charm and youthful curiosity about the world. "His view of happiness is to flit backwards and forwards between Boston and Tokyo," wrote a friend. Lowell spent long stints in the Far East, in spacious homes maintained by servants, where he emerged as one of America's leading interpreters of Japanese culture as he anthropologically chronicled its transformation.

In 1889, while on a Japanese train, Lowell eyed his fellow passengers as if they were scientific specimens. "I would willingly have chloroformed them all, and presented them on pins to some sartorial museum," he later wrote in what he intended as jest, "for each typified a stage in a certain unique process of evolution." The changes in Japanese society could be seen in the eclectic mix of clothing—traditional dress combined with Western styles. "Sometimes a most disreputable Derby, painfully reminiscent of better bygone days, found itself in company with a refined kimono and a spotless cloven sock. Sometimes the metamorphosis embraced the body, and even extended down the legs, but had not yet attacked the feet." It was meant to be a lighthearted observation about a culture he deemed benighted, but it held deep significance.

America and Europe had grown fascinated with foreign lands not only because they were so different from the West but because they offered case studies in what was perceived as a process of cultural evolution. Nineteenth-century anthropologists placed societies on a gradient from "savage" to "barbarian" to "civilized," a ladder of human progress that reflected and reinforced racist attitudes. (Anglo-Saxons inevitably saw themselves at the top of the ladder.) Although Lowell praised aspects of Japanese society—its art, architecture, gardens—he contended that the Far East was evolutionarily stagnant, "half civilized." It was a conclusion that comforted his educated readers in Boston, New York, and London. They could rest assured of their superior status in the hierarchy of humankind.

But even Westerners could acknowledge that they themselves were hardly pinnacles of evolution, that America and Europe had not yet achieved a hoped-for state of cultural perfection. "There is no doubt, one might remark, that the most highly civilized nations of the earth do not know how to conduct themselves," an article in *The New York Herald* conceded, "that their intelligence is exercised chiefly in killing each other, in ruining one another, each for his own profit; that they discount the future like blind men, like fools; that neither thieves nor assassins are rare among them." The article had been written by an astronomer, a Frenchman named Camille Flammarion, who argued that to find a society higher on the evolutionary hierarchy, one should look neither east nor west—but up.

CHAPTER
2

THE FRENCH PHILOSOPHER

1891–1892

As the nineteenth century staggered into its final decade—with Percival Lowell, the American expatriate, luxuriating in the mannered homes of the Far East—Western societies churned under relentless change and disruption. Rapid urbanization and industrialization had so accelerated the pace of life that it appeared, to many in its midst, that the Earth might fly apart from centrifugal force. A gaping divide between workers and elites stoked not just resentment but class warfare. Agitators and terrorists in the United States and Europe were attempting to overthrow the capitalistic system through bombings, strikes, and assassinations. "All are awaiting the birth of a new order of things," declared a prominent anarchist who hoped to midwife that new order. The era came to be known, especially in France, as the fin-de-siècle, a term that literally meant the end of the century but conveyed, as well, a sense of decadence and foreboding. A noted passage in Oscar Wilde's *The Picture of Dorian Gray* gave voice to the mood of the age: "'*Fin de siècle*,' [sic] murmured Lord Henry. '*Fin du globe*,' answered his hostess." To many, it felt like the end of the world.

In this time of trepidation, the larger public grasped for signs of hope and renewal. At a Catholic nursing home in the shadow of the Pyrenees, in the French resort town of Pau, an elderly widow yearned to leave a constructive legacy. Anne-Émilie-Clara Goguet Guzman spent her final years alone. She had lost her husband, a well-to-do industrialist, decades earlier and then had poured herself into raising their son. The boy grew to become a military officer, a strong man with a sensitive side who pondered the meaning of life and the immortality of the soul. When he, too, died young, leaving no family of his own, his grieving and wealthy mother sought to keep his spirit alive. Mme. Guzman was a creative woman—she wrote fairy tales and poetry—and as she watched her own mortal end approach, she applied her imagination to the drafting of her will.

The wider world learned the result in the summer of 1891. "An old lady has just died," a cable dispatch from Paris announced, "leaving 100,000 francs as a prize to the astronomer, French or foreign, who . . . shall be able to communicate with any planet or star." "The prize is to be named the 'Pierre Guzman Prize,' in memory of Mme. Guzman's son," *The Chicago Daily Tribune* elaborated, describing her will as "very curious."

The bequest was certainly surprising, but the founder of the Astronomical Society of France did not consider it odd. He rose to address his colleagues at a meeting that autumn, at the society's offices in Paris's Latin Quarter. "The venerable lady of Pau, Mme. Guzman, may sleep in peace," the astronomer said as he praised her for promoting "the progress of the most beautiful of sciences." As for the origins of the widow's gift, the speaker added, "I'm humbled and flattered to have been the indirect cause."

Camille Flammarion was more than an astronomer; he was a starry-eyed prophet, a man who induced people to think beyond themselves, beyond their lives, beyond the Earth. "Astronomy plunges us into the insoluble mystery of the infinite and of eternity," he declared, "and therein lies its grandeur." A scientist, philosopher, lecturer, journalist, and novelist, Flammarion was among his country's most beloved intellectuals, one who made the wonders of the night sky accessible to all. He imbued the study of the universe with emotion and spirituality, and he cut the figure of a bohemian adventurer. After his wedding in the summer of 1874, he famously took his wife, Sylvie, on a

CAMILLE FLAMMARION.

honeymoon balloon ride by starlight.

Flammarion had long obsessed about alien life. In 1862, at just twenty years of age, while serving as an assistant at the Paris Observatory, he published his first book, *The Plurality of Inhabited Worlds*. It was a thin volume that made an expansive claim. "Life is a law of nature," he later summarized his thesis, "it overflows the entire Earth, like a cup too narrow to contain it, and the other worlds will show us the same when we know how to discover it." In his fiction, he imagined other planets to be populated by the souls of dead humans. One of his novels, unabashedly autobiographical, centered on a French astronomer who goes with his bride-to-be on a nocturnal balloon flight, although in the book the adventure ends tragically when the lovers plunge to the earth below. They soon reunite, reincarnated on Mars.

Flammarion's writings drew a passionate and cultlike following. One fan, a wealthy and aging gentleman, gave Flammarion a château on a hill above the Seine in the Paris suburb of Juvisy. The astronomer converted the house into an observatory dedicated to the study of the planets. Other devotees sent him jewels and artwork, and one sent a gift both morbidly intimate and bizarre. When Flammarion received the package, he noted its soft texture and strange odor. An accompanying note, from a doctor, explained that the parcel contained the offering of a secret admirer—a young countess who had died of tuberculosis:

She begged me to send you, the day after her death, the skin

FLAMMARION'S OBSERVATORY AT JUVISY.

of her lovely shoulders. This is the skin, and you must promise that you will use it to bind a copy of the first book you may publish now, after her death. I have delivered this souvenir to you, Monsieur, as I faithfully promised.

You can still feel Flammarion's pull today on a visit to Juvisy, where his observatory sits like a moldering castle on a busy motorway that leads to Paris-Orly Airport. I stopped by one early April before flying out the following day, with only a few hours to tour the property and explore its collections. In that short time, my primary goal was to gain a visceral sense of the man, to immerse myself in his world.

To access the property, I entered through a metal gate beneath an ornamental archway that Flammarion had inscribed with the motto AD VERITATEM PER SCIENTIAM—"To truth through science"—then found myself in lush terrain that rolled with rises and swales. The trees were newly leaved, the air scented with lilac. In the overgrown hedgerows and thickets I found relics of Flammarion's active life: a meteorological station, now in ruins; a commemorative plaque beside a cedar that had been planted by one of his admirers, the Brazilian emperor Dom Pedro II, in 1887.

I walked through the château itself, past its walls of peeling paint and crumbling plaster, up to the observatory dome, which still houses working equipment. There I was granted the privilege of viewing a sunspot projected onto a screen through Flammarion's own telescope. But the highlight for me was the library—shelf after shelf packed with treasures and bric-a-brac, from statuary to fossils to celestial globes, as well as ornate, gilt-edged books, many written by Flammarion in his long and prolific lifetime. Among the volumes, I spied one in red leather—a copy of *The Plurality of Inhabited Worlds*. The bottom of its cover bore a small imprint in gold. RELIURE EN PEAU HUMAINE, it read. Bound in human skin.

. . .

FLAMMARION'S ALLURING NOTIONS ABOUT extraterrestrial life may have seemed fresh and bold to lay readers in his day, but to scientists these ideas were the discarded beliefs of a bygone era, a time when even the most eminent stargazers naïvely embraced the existence of aliens.

In 1610, the German astronomer Johannes Kepler, famous for his laws of planetary motion, asserted "with the highest degree of probability that Jupiter is inhabited." Some seventy-five years later, an influential scientist of the French Enlightenment, Bernard Le Bovier de Fontenelle—injecting the casual racism of the period—imagined Venus as the home of "little black people, scorch'd with the Sun, witty, full of Fire." The following century in Britain, William Herschel—discoverer of Uranus—believed he saw forests and towns on the moon ("promise not to call me a Lunatic," he begged a colleague), then further argued that "the sun is richly stored with inhabitants." Attitudes in America were hardly more sober. Benjamin Franklin scribed, "It is the opinion of all the modern philosophers and mathematicians, that the planets are habitable worlds." By the early nineteenth century, one prominent author and scientist confidently estimated the total population of the solar system at almost 22 trillion—a tally that included not just those living on the planets but also the imagined residents of the asteroids and even the rings of Saturn.

This notion that life was all-pervasive in the universe, however, soon faded. New observations and theories showed that the smaller celestial bodies, up to and including Earth's moon, retained no appreciable atmosphere—providing no air to breathe—while the solar system's largest planet, Jupiter, was of such low density that it likely possessed no solid surface. The outermost worlds appeared to receive so little light and heat that they were probably too cold for intelligent life, and those closest to the sun—Venus and especially Mercury—were too hot. The default position of astronomy now flipped; scientists assumed that life was absent elsewhere unless strong evidence could be found to the contrary. The conditions on other planets were simply too dissimilar from those on Earth to support earthlike beings.

"Well! This reasoning of savants is the reasoning of fishes," Flammarion shot back at the skeptics, whom he accused of narrowmindedness. He offered a parable:

> Imagine for a moment two silvery [minnows] at the bottom of a river flooded with sunlight talking together. Fishes understand one another very well, despite their apparent muteness.
>
> One of them, who has more than once been nearly taken by the net, but who, with a certain cunning, recognizes fishermen at once, assures his comrade that there is a world outside of the water.
>
> This [other] one, perfectly familiar with the conditions of fish life and with the functions of the gills, easily crushes his enlightened adversary with the weight of his scientific arguments. "Live out of the water! What sensible fish would believe such chatter for a moment! Come, now! An oyster would not believe such tales. The shadows that we see passing on the banks cannot be living. They are optical illusions. Live out of the water! What a joke!"

Flammarion's clever prose and enchanting ideas had long seduced readers in Europe. In 1890, his fame leapt the Atlantic when *The New York Herald* began publishing and syndicating his columns in the United States. The Frenchman shared his knowledge of all things scientific—comets and sunspots, meteor showers and electrical storms—in articles translated into English, and in January of 1892 he offered a provocative essay that ran under a tantalizing headline: HOW TO TALK WITH THE FOLKS ON MARS. "A very aged lady, Mme. Guzman . . . had been deeply interested, especially during her last years, in the descriptions of the planet Mars which I have given in my works," Flammarion wrote. He reminded readers of the widow's posthumous gift and insisted that her idea of reaching out to other worlds was "not at all absurd." In fact, he outlined a scheme for doing just that.

Like the Koreans who sent messages with signal fires from mountaintop to mountaintop, two worlds could shout hello across the void using patterns of light and a universal language: geometry. By means of electrical illumination, the people of Earth could trace enormous figures on the landscape—perhaps a hundred miles across, so large they would be visible through a telescope on Mars. Earthlings would then watch for the people of Mars to do the same. "They show us a triangle, we produce it here. They trace a circle, we imitate it," Flammarion explained. "And lo! communication is established between the heavens and the earth for the first time since the beginning of the world." The French astronomer acknowledged that to accomplish this plan required several stipulations be met, beginning with the most basic: "First, that Mars is inhabited."

"All this is bosh and nonsense to me, and I dare say it was bosh and nonsense to the man who wheedled this French woman into setting aside her 100,000 francs," fumed George Davidson, an American astronomer and former president of the California Academy of Sciences, who expressed a sentiment widely shared by his colleagues. Yet even Flammarion's critics agreed that if any world beyond Earth were inhabited, it would be the one that glittered like a sunlit drop of blood among the stars.

. . .

MARS HAS HELD A special place in the public's imagination since ancient times. When sky watchers first identified a handful of celestial lights that drifted mysteriously across the firmament—objects that the Greeks called *planetes*, or "wanderers"—people noted that one glowed a peculiar ruddy hue. Not only was the color unusual but the planet's brightness proved odd—wildly inconstant, at times so dim that it got lost among the stars, at others so brilliant that it glowed like an ember. A further enigma was the planet's path. Rather than transiting the constellations in a continuous direction, it occasionally doubled back on itself before resuming its course. To many cultures of the past—the Babylonians, Persians, Romans—this fiery wanderer seemed an omen or a deity, a representation of death and war.

In more modern times, astronomers would discover that Mars was hardly a god but rather a world like our own, in fact our closest neighbor toward the outside of the solar system. Scientists found that its variable brightness and retrograde motion could be explained by the manner in which that planet and ours orbit the sun. Both move in the same direction, like racehorses on concentric tracks, Earth galloping considerably faster than its laggardly competitor. During most of our circuit through the heavens, Mars is either far behind or far ahead—and therefore dim and distant—but once every twenty-six months we pull alongside and then sprint past. At this time of passage (known as an *opposition*, since it is then that Mars sits opposite the sun in our sky), the red planet appears to stall in its motion and then briefly turn backward while it simultaneously flares in brilliance due to its proximity.

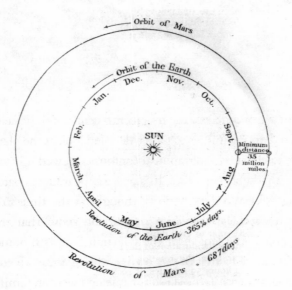

DIFFERENCE BETWEEN THE ORBIT OF MARS AND THAT OF THE EARTH.

The Martian orbit is not circular, however, but broadly elliptical. If Earth passes by when Mars is at the closer end of its sweep around the sun, the two planets will come especially near, making Mars even brighter and bigger from our perspective. This happens once or twice every fifteen years, and the effect is obvious to the naked eye. "Nothing can exceed the calm, fierce, golden, glistening domination of Mars over all the stars in the sky," Walt Whitman wrote in 1879 at just such a time of close planetary opposition.

The effect is even more impressive in a telescope. Through an eyepiece, Mars at opposition looks like a miniature full moon, bright and

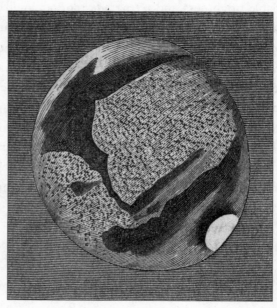

ASPECT OF MARS, WITH ITS CAP OF POLAR SNOW.

round and large. For astronomers of the nineteenth century, these celestial events were prized opportunities to gain a detailed look at the planet next door, a chance to chart its geographical contours. Viewed up close, the planet seemed remarkably earthlike. Its surface appeared as a patchwork of contrasting light and dark regions, thought at the time to be continents and oceans, with caps of white (presumably snow) that grew and shrank at the poles. The length of the Martian day, astronomers knew, was almost identical to that of our own, and while the Martian year was twice as long as Earth's, the seasons came and went in familiar fashion. "[Mars] offers so many points of resemblance to our world," Flammarion wrote, "that we would scarcely feel like exiles if we were to pack up our household goods and transport them thither."

As telescopes improved in the late nineteenth century, astronomers were eager to turn their latest instruments on Mars at each subsequent opposition in an attempt to solve the planet's mysteries. Just such an opportunity was set to arise in the summer of 1892. In early August, Earth would make its closest approach to Mars in fifteen years.

...

THE PEOPLE OF EARTH, meanwhile, focused on pressing matters at home as a wave of violence and unrest washed across Western countries. That spring of 1892, terrorist bombings struck Paris, producing panic and the swift guillotining of one of the perpetrators. In Madrid, anarchists schemed to blow up the royal palace and the houses of parliament until their plots were foiled by police. The United States was no different. With memories of Chicago's 1886 Haymarket riot still fresh in the public mind, now—six years later—bloody conflict erupted at Andrew Carnegie's Homestead Steel Works in Western Pennsylvania, as hundreds of armed guards battled thousands of exploited and locked-out workers and their families. The brutal confrontation killed ten, wounded many more, and risked stalling an enormous and symbolic construction project that relied on steel—the Chicago World's Fair, to commemorate Columbus's arrival in the Americas in 1492.

Meanwhile, night by night, Mars brightened as it drifted across the zodiac—through Scorpio and Sagittarius before settling into Capricorn. At observatories in Europe, the United States, and South America, scientists watched as the waxing gibbous planet grew in their telescopes. Colors and shapes painted the Martian terrain: broad expanses of gray that mingled with ocher; regions of brick red and sea green; here and there a sash of yellow or a patch of chocolate; at the pole, an oblong cap of white. "In the 400th year from Columbus, we feel as if we too were almost discovering a new country," an astronomer at Harvard's Southern Hemisphere observatory in Peru wrote his mother.

Of all the world's astronomical institutions, the one best equipped

THE LICK OBSERVATORY, MOUNT HAMILTON, CALIFORNIA.

to discover the truth about Mars sat in Northern California, high in the Diablo Range overlooking San Francisco Bay. The Lick Observatory was the most expensive and modern facility of its kind. It had opened a few years earlier to great fanfare and now stood atop Mount Hamilton as a gleaming monument to the generosity and ego of James Lick, an eccentric businessman and land speculator who had amassed his fortune during the Gold Rush. The benefactor's corpse lay entombed within the dome, beneath what was—indeed what *had* to be, per the terms of his gift—a telescope "superior to and more powerful than any . . . yet made." With its unprecedented vision, the Lick was figuratively poised to bring Mars closer to Earth than ever before in the history of humankind. Throughout July, the Lick astronomers watched Mars every clear night.

Then, as August began, America's dailies exploded with news, a frenzy of science and hype that would later be recalled as "the Mars *boom* of 1892." The combustible mixture was made especially volatile by a revolution under way in the American news industry. Joseph Pulitzer had recently purchased the near moribund New York *World* and turned it into one of the most successful dailies in the land by employing sensationalistic tactics that were soon imitated by others. Pulitzer focused on titillating subjects and used strategic puffery and embellishment to clothe inconsequential stories in irresistible garb. "I wanted to put into each issue something that would arouse curiosity and make people want to buy the paper," Pulitzer acknowledged. That first week of August in 1892, his editors seized on Mars. Any

New Yorker passing a newsstand or riding a commuter train could hardly have missed the headlines atop the front page on four consecutive days, from COME, VISIT MARS! on Tuesday to MARS AND ITS MEN on Friday.

The paper teased its astonishing news in the subheads. REMARKABLE REVELATIONS MADE BY THE LICK INSTRUMENT ON CALIFORNIA'S MOUNTAINS—STRANGE ILLUMINATIONS WHICH HAVE NEVER BEEN SEEN BEFORE. On the very rim of Mars, where the planet's rotation turned its dayside to night, astronomers had noticed the sudden appearance of three bright spots, like powerful searchlights shining from high peaks set far apart. "Steadily and brilliantly these three effulgent lights glistened like diamonds upon the southwestern limb of the planet, growing more brilliant as the dark purple shadows stole about them or surrounded the mountains upon which they seemed to be placed," *The World* reported. "Were these such lights as Flammarion suggested the people of Earth ought to show to attract attention of those on Mars?" The paper had no answer, but it assured its readers that the apparitions "were no phantasmagoria of the astronomical imagination. They were seen from the Lick Observatory four nights in succession . . . forming a perfect triangle."

CHAPTER
3

What the Colorblind Astronomer Saw

1892–1893

Percival Lowell might well have missed the Mars boom of 1892 had Theodore Roosevelt's cousin not been late for a morning train the previous summer. Alfred Roosevelt, a millionaire Wall Street banker, had rented a beach house for the season on the Westchester side of Long Island Sound, a cool refuge for his family while he commuted into the sweltering city. On a Thursday in early July immediately prior to the Independence Day weekend, Roosevelt's carriage arrived at Mamaroneck Station just as the 8:14 A.M. train to Manhattan prepared to depart. He raced up the stairs to the platform, grabbed a handrail on one of the cars, aimed his foot toward the step—and missed. Roosevelt clenched the rail and struggled to board, but the locomotive had begun to chug forward, dragging him along and throwing him against an iron abutment. As Roosevelt fell, his right leg slid onto the tracks, and by the time the train screeched to an emergency stop, his ankle lay crushed beneath the wheels and blood poured from his scalp.

"Three children remain to comfort the widow of the young banker," *The New York Times* reported on July 4 after doctors' efforts proved inadequate to save his life. The grieving Mrs. Roosevelt, *The New York Herald* reminded its readers, was herself the product of a prominent lineage, for she was the former Miss Katharine Lowell—"of the old Lowell family, of Boston."

Word of his sister's tragic loss reached Percival Lowell in Tokyo, where he was now studying Japanese "occult" practices: firewalking, spirit possession, hypnotic trances. Relatives urged him to return home. "I shall out of regard for the family's wishes in the matter sail on Oct. 22," he wrote his cousin William Lowell Putnam (who happened also to be a brother-in-law, for he had married Percival's sister Elizabeth). Arriving in Boston weeks later, Lowell eased back into the comforts of elite society. He resumed polo, attended cultural exhibitions, graced dinner dances. In January 1892, he was seen at "the event of the season," a performance by the flamboyant actress Sarah Bernhardt at the Tremont Theatre. Yet the focus of his intellectual attention began a subtle shift toward matters scientific. He could be found at meetings of the Boston Society of Natural History, where he reconnected with Edward S. Morse, the zoologist whose lectures on Japan a decade earlier had proved so life-changing.

Lowell had recently read an ambitious work called *The Philosophical Basis of Evolution* by the Scottish astronomer and climatologist James Croll. Although Lowell found its contents "vacuous," the book inspired him to contemplate producing his own scientific treatise on an epic theme. "I have thought of some things to be said about the philosophy of the cosmos," Lowell revealed with grandiosity, though exactly what corner of this vast realm he would visit was unclear. Spring transitioned into languid summer as August's stifling humidity descended. "The miseries of the dog-days are here in full force," the *Boston Evening Transcript* lamented. It was then that the Mars boom sounded.

...

NEWS OF THE TRIANGLE on Mars, hyped by the New York *World* and quickly copied by other papers, ricocheted across the land.

IS THE PLANET INHABITED BY PEOPLE TRYING TO SIGNAL THE EARTH, a South Dakota headline asked. A West Virginia daily seemed to reply: CONCLUSIVE EVIDENCE THAT THE PLANET IS INHABITED AND

THAT ITS PEOPLE ARE HIGHLY CIVILIZED. The New York *Press* assessed the evidence with more equanimity. "It is not impossible," the paper editorialized, "that the three great lights seen through the Lick telescope for some nights past . . . may be artificial; that the people of Mars, knowing that this is their best opportunity for fifteen years, are trying to signal to the people of earth."

It was, in all seriousness, a silly interpretation undermined by two simple arguments: First, the shape traced by the lights proved hardly significant; any three points not perfectly aligned will, of necessity, form a triangle. Second, the astronomers who had witnessed the phenomenon believed there was a natural cause; they assumed the bright spots were due to the sun glinting off snowcapped peaks or off clouds along the planet's darkened edge. Even Camille Flammarion considered this natural explanation likely.

The public, well inured to the sensationalism of certain news outlets, read the stories as light entertainment, a diversion to quench the heat of August. "The latest astronomical news has equally interested Mr. and Mrs. Jones, and they have been able to discuss it in the most amiable—almost jovial—style with Mr. and Mrs. Smith," a New York magazine related. "Even the children . . . are, perhaps, quite as intelligent in their speculations as their unusually condescending elders. At the breakfast-table, along with the iced cantaloupe and Mary Jane's zealousness, we have also devoured everything in the shape of additional information or disputation that the morning paper has offered [regarding] . . . these small lights on the left limb of Mars."

But for all the talk of the bright triangle on Mars, of greater consequence were other shapes on the planet: dark lines that scored the surface like claw marks. Their meaning, as well as the question of their very existence, had vexed astronomy for fifteen years now, ever since they had first been reported by an Italian observer on the last occasion when the red planet sailed so close.

. . .

ON A MILD MORNING beneath a sky of broken blue, I arrived from Paris in Milan. Debarking my train into the city, I navigated a warren of cobbled streets to the Palazzo di Brera, a late-Baroque palace of archways and brick that surrounds a central courtyard. I encountered a towering bronze statue, a heroic nude with a paradoxical name: *Napoleon as Mars the Peacemaker.*

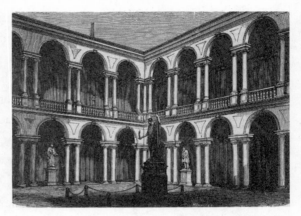

MILAN: COURTYARD OF THE PALAZZO DI BRERA.

As an interpretive sign explained, the legendary French emperor, as conquerer of Italy in the early nineteenth century, ordered his army to assemble paintings here for public display and thereby created one of the great art museums of Europe, the Pinacoteca di Brera. It remains an immense lure for tourists and was the reason for the crush of people around me. I, however, carved off from the crowd and wound my way to the back of the complex, then climbed several flights to a small astronomical museum.

I had arranged to meet a local host here, and together we ascended farther, to the top of the building, and exited onto a narrow metal walk that followed the apex of the long roof. Toward the south I caught the spires of Milan's famed cathedral, and to the north, I was told, one can see the High Alps on a clear day, but we progressed straight ahead to reach what looked like a gargantuan oil drum positioned atop the palace. It was the copper dome of the historic Brera Observatory, and it housed a consequential telescope: a ten-foot tube of wood and brass set upon an angled cast-iron base. Various gears and knobs enabled the user to direct the glass toward any desired object in the sky and to hold the image fixed even as the Earth turned below one's feet. This was made possible by a clock drive, like the innards of a timepiece, which moved the telescope infinitesimally to counteract our planet's rotation. My host demonstrated by winding it with a key, raising lead weights on a wire that then pulled slowly downward. The

GIOVANNI SCHIAPARELLI.

mechanism ticked like a grandfather clock, marking the slow passage of time, while I projected my imagination back a century and a half to the originating event that had caused me to come here.

It was the evening of August 23, 1877. The Brera Observatory's director, Giovanni Schiaparelli, had climbed to this same perch to use this same telescope. He was forty-two years old and already a scientist of international renown, for he had famously explained the cause of meteor showers when he discovered that shooting stars rain down due to the Earth's passage through debris shed by comets.

In his personal life, Schiaparelli could be absentminded—soon after his wedding, he inadvertently abandoned his wife at a hotel, then returned with an apology: "I completely forgot I was married!"—but in his work he was thorough and dedicated. For him, preparing for a night of observing required the forbearance of an athlete before a marathon. "I have adopted the rule to abstain . . . from everything which could affect the nervous system," he once remarked, "from narcotics and alcohol, and especially from the abuse of coffee."

What had brought Schiaparelli to the rooftop that night in 1877 was a total eclipse of the moon, but that did not hold his interest for long. Mars was then making an extremely close approach to Earth, and the red planet, blazing in the sky, beckoned. Schiaparelli turned his telescope toward our neighboring world and began what would become a campaign of regular observation that continued for months, his goal to create a detailed map of the Martian surface. The Italian astronomer saw Mars in ways others did not. He was colorblind, a condition that deprived him of the ability to detect subtle gradations in hue but may well have enhanced his perception of shape and con-

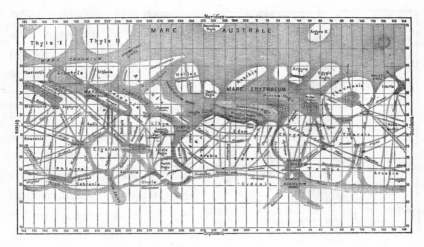

SCHIAPARELLI'S MAP OF MARS.

trast. Schiaparelli had originally trained as an engineer and architect, and he brought to his task a draftsman's sensibility; he imposed on Mars its own latitudes and longitudes and plotted its features with trigonometric rigor, as if conducting a land survey. Through the fall of 1877 and the subsequent winter, he constructed his map. Every two years after that, when Mars again swung near at opposition, he refined it. The results baffled the astronomical community.

Schiaparelli depicted not merely the dark and light regions that others saw—the features generally referred to as oceans and continents—but also an abundance of narrow streaks that appeared to connect the seas one to another. These delicate stripes extended for hundreds, even thousands, of miles with nary a bend, so straight and slender that they seemed to run with a sense of purpose. The lines intersected, often many at the same spot. And although their position held fast, their character changed through the years and seasons, even week to week. "They may disappear wholly, or be nebulous or indistinct, or be so strongly marked as a pen line," Schiaparelli explained. Most bizarre, some of the lines would occasionally double, as if suddenly paralleled by a twin, a phenomenon he called *gemination*. "What could all this mean?" Schiaparelli asked the Belgian astronomer François Terby, who had made his own extensive drawings of Mars.

Since the strange linear features connected what were assumed to be oceans and looked superficially like stretches of water, Schiaparelli called them *canali*. In Italian, the word meant "channels," but when it found its way to the English-speaking world, the term was promptly mistranslated. CANALS ON THE PLANET MARS read a headline in *The Times* of London in 1882. The term stuck.

No one of note seriously believed that the "canals" were waterways of artificial construction, but whatever they were, astronomers across the globe headed to their telescopes to see if they, too, could spot the enigmatic features. The results proved inconclusive. In the years when Mars again came close to Earth—1884, 1886, 1888—a handful of scientists said they saw the fine lines, yet most saw none (or were reluctant to admit that they had observed the ghostly markings for fear of being ridiculed, a condition that Schiaparelli's Belgian friend Terby titled "Schiaparellophobomania"). In 1890, an American astronomer concluded, "It is perhaps hardly yet safe to regard their reality as fully established." He then added, "In 1892 the planet will again be favorably situated for observation, and we may reasonably expect that then the doubts will be resolved." Indeed, in that summer of 1892, while news swirled about the triangle on Mars, confirmation seemed to arrive of Schiaparelli's lines on the planet. They were seen at Flammarion's observatory in France, Harvard's in Peru, the Lick in California. Again, the boom rang out.

"What is going on in Mars?" *The New York Herald* asked. "Are the so-called 'canals' really signals which are being exhibited to us, or are they made to connect all the big seas one with another?" The New York *World*, citing an idea of Flammarion's, suggested the lines might have been laid out as "a gigantic semaphore . . . which the Marsians were hanging out for people on the earth to see and answer." *The Boston Daily Globe* speculated further: "Who shall say that some day a delegation of Marsonians will not visit the earth, and perchance take a peep through a telescope even larger than the Lick instrument, and gaze delightfully at the planet of their birth." ("Marsians," "Marsonians," "Marsites," "Martials" . . . many fanciful names were given to the planet's hypothetical inhabitants before "Martians" emerged as the consensus term.) The madness in the press became too much for astronomers to bear.

"Pshaw!" an observatory director in Oakland, California, huffed. "Do you know that every true astronomer has been very sorry to see all this haberdash that some of the newspapers have been printing about Mars being inhabited and our seeing canals on the surface, and all that other rot and nonsense. Fairy tales and wild speculations have no place in a serious science like astronomy."

. . .

OF ALL THE SCIENTISTS irritated by the talk of signal lights and artificial canals on Mars, none was more disdainful than Edward S. Holden, head of the Lick Observatory. A brusque man of military training who was detested by his own staff (his assistants mocked him as "the Czar" and "the Dictator"), Holden turned his scorn on newspapermen, and on Flammarion, for promoting such fanciful ideas. As if emulating Zeus, who hurled thunderbolts from Olympus, Holden flung barbs from Mount Hamilton. "All, or nearly all, of the present excitement over Mars is merely exaggeration and a sham excitement," he vented in a telephone call to the Associated Press. To *The New York Herald* he fired a telegram: "The prevalent excitement on the subject of Mars is a manufactured one for which astronomers are not in the least responsible. . . . Our simple duty is to tell the simple truth and not to lead ourselves to any sensation." In a sit-down interview with the San Francisco *Examiner*, he underscored, "I am by no means ready to say that these so-called canals are veritably filled with water, and I do not believe that they are the work of human hands. If you ask me what these dark markings really are, I am obliged to answer that I do not know."

"'I do not know!'" screamed an exasperated Flammarion. "One could make the same reply to every imaginable question":

> "What is the human skull?" "I do not know."
> "What is the nerve system?" "I do not know."
> "What is the origin of man?" "I do not know."
> "How is it that the sun heats us?" "I do not know."
>
> It seems to me that if we were always content with this answer humanity would still be in the age of carved stone and caverns inhabited by rhinoceroses and bears.

Flammarion argued that it was the scientist's job not simply to recognize the limits of knowledge but to push further—to creatively explain the unexplained.

The divide between the two men revealed a chasm that split the broader astronomical community, a rift born out of differences in sentiment and outlook. On one side of the line huddled cautious researchers who considered it their duty to remain unemotional, precise, and mathematical—to collect data and do little more. On the opposing side were more broadminded scientists—many of them also science popularizers—who argued that genius lay in the *interpretation* of data, who saw the known as a mere starting point for envisioning what may be. The schism took on the aspect of an intractable political fight. "There are [conservatives] even in science who regard imagination as a faculty to be feared and avoided rather than employed," complained the Irish physicist John Tyndall, who allied himself with the more liberal wing. "Bounded and conditioned by cooperant Reason, imagination becomes the mightiest instrument of the physical discoverer. Newton's passage from a falling apple to a falling moon"—that is, the discovery of the universality of gravitation—"was a leap of the imagination."

Percival Lowell, though not yet a working astronomer himself, clearly fell into this latter camp. "Imagination is the soul of science," he once had told an audience in Japan. "Certain it is that there is quite as much imagination required in the pursuit of physics as of poetry."

IN OCTOBER 1892, SCIENTIFIC imagination went on display in Boston, at the Tremont Theatre, the same venue where Sarah Bernhardt had seduced Lowell and the city's elite back in January. The playhouse now presented an unusual spectacle that merged the artistry of grand opera with the wonders of astronomy. During the two-hour matinee, eager theatergoers "imagined themselves for a time to be floating in space," one reviewer noted. "There has seldom been in Boston an audience more truly representative of the culture of the city than that which gazed at the strange scenery of 'A Trip to the Moon.'"

This theatrical extravaganza created the illusion of spaceflight through elaborate stage sets, slide projections, the clever use of scrims and drops and gauze, and more than eight hundred incandescent bulbs that simulated the lighting effects of sun and stars from shifting angles. At the far end of the voyage, the audience landed on the moon and was offered a wholly alien perspective—looking back from the cratered surface (made of painted canvas) toward a pasteboard Earth turned by a hidden hand crank. "Slowly as time goes by the globe revolves on its own orbit through this black and depthless sky, Europe and Asia fade away, America looms up in the west," a viewer described this awe-inspiring moment as performed in San Francisco. "People hold their breath when they watch this scene, the extraordinary inversion of nature as we see it amazes them."

Narrating from the stage and playing tour guide to the cosmos was "a gentleman long and favorably known for his popular articles on scientific subjects." This was one Garrett P. Serviss, a bespectacled writer who, despite his bookish demeanor and meager build, spoke with confidence and brio. "Looking at such a landscape as this, we are irresist-

GARRETT P. SERVISS.

ibly impelled to inquire, are there any inhabitants on the moon fitted to enjoy these scenes? Evidently we must reply no," he told his audiences, "for without air and water we cannot conceive of the existence of beings resembling the inhabitants of the Earth, and if we begin to speculate on the existence of creatures requiring no air and water for their support, we must at once abandon all scientific ground."

As an author, columnist, and lecturer, Serviss was as influential as he was prolific, a voice one could trust in this age of sensationalistic journalism. "In his use of imagination Mr. Serviss never perverts truth, and that is one secret of his success," his hometown paper, *The Brooklyn Daily Eagle*, chauvinistically asserted. Given his popularity and respect, Serviss proved the ideal pundit to set the record straight on controversial matters in astronomy, including the bewildering array of claims about Mars.

"The results of the observations made [this year], so far as they have become known, indicate a substantial verification of Schiaparelli's discovery that Mars is covered with a curious network of lines or narrow bands," Serviss had written in *Harper's Weekly*. He clarified in the New York *Sun*, however, that "no reputable astronomer would give countenance to the idea that these lines . . . are artificial canals" and that "no discovery has yet been announced which justifies us in asserting positively that living creatures exist upon Mars."

What, then, might those "canals" represent? Scientists had offered various theories: Perhaps they were rivers or mountain chains, volcanic fissures or glacial crevasses, or the tracks of meteors that had grazed the planet's surface. "But nothing has yet been proposed that covers all the appearances presented," Serviss wrote in *The Popular Science Monthly*. In other words, no credible and cohesive theory could explain how the canals followed exceptionally straight courses, tended to meet at common intersections, sometimes appeared double, and often appeared not at all. "That there is something very singular on that planet no one can doubt who has looked at it with a powerful telescope," Serviss concluded. The scientific world now needed someone with the intelligence and imagination to fit the pieces together.

...

IN THE WANING MONTHS of 1892, Mars once again receded into the distance, but Percival Lowell's mind could not move on. He remembered gazing at Mars through a small telescope as a child, and now—in his late thirties and possessing a larger telescope—he penned a letter to the director of Harvard's astronomical observatory. "Could you kindly tell me," he inquired, "[what are] the most modern charts or drawings of Mars? . . . Are Schiaparelli's, Terby's, procurable?"

That December, Lowell departed on what would be his final trip to Japan, and he brought the telescope with him. As the new year began, Mars was no longer in good position for viewing, but Lowell explored the solar system. "Last night I observed Saturn," he wrote his cousin (and brother-in-law) William Lowell Putnam from Tokyo in the winter of 1893. "I shall probably not be back [in Boston] till the middle or late autumn." Home by Christmas, Lowell received a holiday gift from his aunt. It was a thick book, imported from abroad, titled *La Planète Mars et Ses Conditions d'Habitabilité* (*The Planet Mars and Its Conditions of Habitability*). The author was Camille Flammarion. Lowell, fluent in French as befitted a man of his station, soon lost himself in those six hundred pages.

When an endeavor struck Lowell's fancy, he was not one to delay but would launch into it abruptly, declaring: "At once!" In this case, after immersing himself in Flammarion's prose, he took up a pencil, opened the front cover of the book, and wrote inside a cryptic note in large cursive script. It said, simply, "Hurry."

CHAPTER
4

Mars Hill

1894

What Percival Lowell had read in Camille Flammarion's book became his own perceived reality less than six months later, when in late spring of 1894 he began exploring Mars via a new and enormous telescope. Through the tube, which stretched twenty-six feet toward the heavens, the planet resembled the pasteboard Earth seen on the stage of Boston's Tremont Theatre. The illumined orb hung against the black curtain of space and rotated unhurriedly, revealing new terrain hour by hour, night by night. Here on display were some of the landforms Lowell had seen on Giovanni Schiaparelli's maps, places the Italian astronomer had labeled whimsically with names plucked from the Bible and ancient legend. Eden. Atlantis. Utopia. The Fountain of Youth. Schiaparelli called his Mars a jumble of "mythical and poetic geography." For Lowell, gazing through a telescope, it must have felt like visiting a land from a fairy tale and discovering that the fables had all been true.

Lowell embarked on his interplanetary journeys from a domed observatory of wood and canvas that sat on a high slope in a pine forest in the American Southwest, far from his New England home. Serenaded by the occasional howl of a coyote, "like the wail of a lost soul," he studied Mars with hypnotic intensity, then sketched and described what he saw. At this time of this year, Mars was most easily observed in the small hours of the morning, around dawn. "Seeing growing dim owing to the growing day," Lowell wrote in his logbook shortly after sunrise on June 7. He then realized the cause of the dimming was not the sun but his own beginner's incompetence, for he had failed to keep the slot in the roof above him aligned with the telescope. "Ha! ha! not day but dome in the way!"

At thirty-nine, Lowell was intent on transforming himself from a man of literature to one of science—and quickly—his hurry due to the unceasing waltz of the planets. In four short months, Earth would once again lap Mars, bringing the two worlds almost as close as they had been during the boom of 1892, as if at hailing distance. By the time of this astronomical opposition, Lowell aimed to acquire the skills to help solve the red planet's mysteries. For the moment, as Mars approached, he familiarized himself with its terrain. He often gazed at the great expanse of white around the South Pole, which sat atop the planet in the telescope's upside-down view. His eye caught bright flashes—sun glinting off sheets of ice, he presumed—a sight he interpreted poetically, as though a message from another world meant just for him. He calculated that it took nine minutes for those rays of light to make their journey across space, "and, after their travel of one hundred millions of miles, [they] found to note them but one watcher"—Lowell himself—"alone on a hill-top with the dawn."

Yet as enchanted as he was by the sparkling snows of the polar cap, what he most sought—what he strained to perceive—were the strangest features of all. "With the best will in the world," he wrote in his logbook, "I can certainly see no canals."

...

LOWELL HAD ENDED UP at this lonely observing post, on a hilltop in Arizona, through a sequence of events that began in Boston with a pair of longtime acquaintances.

Edward C. Pickering, one of America's most powerful and respected scientists, was director of the Harvard College Observatory. William H. Pickering, Edward's younger sibling by a dozen years, was also an astronomer on the observatory's staff. William was captivated by Mars, and in 1893 he proposed to his brother that Harvard launch an expedition to study the planet when it again neared the Earth in 1894. His plan was to erect a large telescope in Arizona, where the dry air would offer especially good views. William Pickering appealed to Harvard for funds, but the university maintained that it could ill afford the thousands of dollars required. So he looked elsewhere for money. He applied for grant support and approached philanthropists. "I find it very difficult to raise the money," he confessed. "Times are hard."

Times were, in fact, abysmal. A Wall Street crash—the Panic of 1893—had triggered economic depression. "Mills, factories, furnaces, mines nearly everywhere shut down in large numbers," a New York financial weekly said of the carnage. William Pickering needed the support of someone whose wealth ran so deep that it could withstand the economic storm. At a dinner party in Boston, he found his angel. "Mr. Percival Lowell . . . now comes to the rescue and provides the necessary means for starting the Arizona expedition," *The Boston Herald* reported in early 1894. "This donation of Mr. Lowell came like a benediction."

It also came with strings. Lowell, who was interested in turning his energies toward science, insisted that if he were to provide the money, he must be in charge. So instead of funding an expedition on behalf of Harvard, Lowell hired William Pickering away from Harvard to execute the plan on his own behalf. Pickering took a leave of absence from the university and also brought to the venture an assistant, a rising young astronomer named Andrew Ellicott Douglass.

To A. E. Douglass fell the first important task: scouting for a promising observatory site. At the end of February, he departed Boston for Arizona—at the time a territory, not yet a state. Its terrain had been controlled for centuries by a shifting mosaic of Indigenous and Hispanic cultures but now fell under the firm grip of Washington, D.C. Less than a decade earlier, the Apache leader Geronimo had surrendered to the U.S. Army, bringing to a close the military's cruel campaign of bloody massacres and forced displacement to drive out the native inhabitants for the benefit of ranchers, miners, and moneyed East Coast investors. Douglass, the astronomer, arrived in the region in early March 1894 and carried with him two coffinlike boxes that held the components of Lowell's "far travelled telescope," the one that had been to Japan and back. Douglass used the instrument to test the clarity of the atmosphere at his various stops.

By train and dusty stagecoach, Douglass toured Arizona's now fading frontier, a landscape later mythologized in Hollywood westerns. He sent Lowell daily updates, including assessments of the towns he visited as places his boss might want to inhabit for a spell. "Tombstone will not be suitable for ladies till a railroad goes through it," Douglass wrote from the boom-and-bust mining town where, in 1881, the legendary lawman Wyatt Earp and his brothers shot down a party of gunslinging cowboys near the O.K. Corral. Moving northwest to a more-settled Tucson, home to a new university, Douglass noted additional drawbacks of a wild nature: "centipedes, scorpions, tarantulas, rattlesnakes and [rabid] skunks." Phoenix, he learned, grew so infernally hot that anyone who can leave in the summer does leave in the summer, and "those who remain sleep out doors at night. They think it would be difficult to sleep during the heat of the day"—as an astronomer must between nights of observing.

Douglass then visited the higher, cooler, forested regions of Northern Arizona. He stopped near the base of the snowcapped San Francisco Peaks, south of the Grand Canyon. "Flagstaff is a town of 1200 to 1500 people . . . in a level valley 2 miles wide," he wrote from the pine-scented community alive with the sounds of sawmills. Flagstaff boasted sapphire days and diamond-studded starry nights that earned it the nickname "Skylight City." Douglass's telescopic tests confirmed that the astronomical observing was good.

FLAGSTAFF, ARIZONA, AND THE SAN FRANCISCO PEAKS.

A mere decade earlier this area had been plagued by gunfights, cattle rustlers, and vigilante justice, but Douglass reported that it was now known for "fair stores and many pleasant people." There were churches and schools, a bank and a library, a baseball team and brass band. An Arizona newspaper explained that the new complexion of the region was due to the coming of the railroad. "Barbarism all along the line has yielded gradually to the advance of civilization."

Lowell liked what he heard. By mid-April, he had decided: "Flagstaff it is."

Douglass promptly began constructing the dome and installed an enormous telescope that was shipped west from Pittsburgh. He placed it on a promontory just west of town, close enough for convenience but far enough to offer a bit of solitude. Douglass was curious how he should refer to his boss's nascent scientific institution when talking to townspeople. "In answer to name," came the reply, "simply call it the Lowell Observatory." As for the observatory site, it, too, received a new and appropriate moniker. The elevated parcel would come to be known as Mars Hill.

THE OBSERVATORY OF MR. PERCIVAL LOWELL OF BOSTON AT FLAGSTAFF, ARIZONA.

...

THAT LOWELL, A MAN with no professional experience or advanced training in science, would abruptly establish his own astronomical observatory may seem an act of utter audaciousness, and perhaps it was. It certainly was newsworthy—"a gentleman long and favorably known as a Japanese scholar, has suddenly appeared as an amateur astronomer of the strongest type," a Boston paper noted—but Lowell's metamorphosis was hardly unprecedented. Many respected stargazers in the nineteenth century were amateurs—merchants, doctors, lawyers, and others who in later life devoted their spare time to watching the sky.

By 1894, the field of astronomy was professionalizing, and amateurs had an increasingly difficult time competing with well-endowed institutions like California's Lick Observatory. Lowell, however, aimed to do just that. With his considerable means, he acquired some of the best equipment available, and he had already hired those two Harvard astronomers, Pickering and Douglass, to assist. An additional hallmark of the Lowell Observatory that distinguished it from other amateur outfits—and from professional ones too—was its primary purpose.

"With regard to the observatory's plan of work," Lowell told an audience in Boston just before he left for Arizona at the end of May 1894, "[it] may be put popularly as an investigation into the condition of life in other worlds, including last but not least their habitability by beings like if unlike man. This is not the chimerical search some suppose. On the contrary," he insisted, "there is strong reason to believe that we are on the eve of pretty definite discovery in the matter." During his talk, Lowell displayed to his audience a globe that was oddly colored, like a pumpkin. This was Mars, he explained—the main focus of his interest. He pointed to the strange lines on its surface, Schiaparelli's canals. "Speculation has been singularly fruitful as to what these markings . . . may mean," he said. "Each astronomer holds a different pet theory on the subject, and pooh-poohs those of all the others. Nevertheless, the most self-evident explanation from the markings themselves is probably the true one; namely, that in them we are looking upon the result of the work of some sort of intelligent beings."

A Boston newspaper printed Lowell's bold predictions, and a clipping of the article soon found its way to California, to that self-appointed arbiter of astronomical propriety, Edward Holden, the Lick Observatory's director. "The foregoing words seem to me to be especially misleading," Holden wrote in a lengthy rebuttal dripping with condescension, which he published in a scientific journal that he himself controlled.

"Nearly every living astronomer will agree with me in saying, as I do, that there is no reasonable probability whatever of any such settlement [of the question of Mars's habitation] at the present time."

Perhaps proud that he was already stirring attention in his new field of study, Lowell mailed a copy of Holden's critique to his family. His mother answered. "Holden seems like a fly that only comes back to the same place to tickle and rouse you again," she wrote consolingly to her son, who was on a train to Flagstaff. "Good-bye my darling may you discover more problems in Mars!"

...

LOWELL, THE FASHIONABLE BACHELOR who had once found contentment flitting back and forth between Boston and Tokyo, now enjoyed splitting his time between New England and Arizona. He came to appreciate the rough, pure existence of the manly West— the wild scenery, the plainspoken people, even "the young Flagstaff band that is learning to play in tune." Lowell trod the dirt streets in evening darkness as he walked the mile from his hotel downtown to the telescope on Mars Hill. He and his assistants divided the night into watches, taking turns at the eyepiece and occasionally sleeping in a tent within the dome. (Later, Lowell would build a cottage, then a house, beside the observatory.) Curious townsfolk and tourists occasionally interrupted the routine. "Are you observatory people?" a voice called out from a passing carriage one August eve as the astronomers ascended the hill. Lowell responded affirmatively and cordially, inviting the vehicle's passengers up for a view of the stars. His guests marveled as they peered through the glass. "Just like diamonds," they effused.

Lowell's own gaze remained fixed on Mars, and as the weeks passed, he watched a drama unfold. When he began his observations, in June, it was early spring in the Martian southern hemisphere, which tilted favorably toward Earth. As the season progressed, Lowell traced the effects of the clement weather. The white expanse around the South Pole, which at first so impressed him with its bright sparkles, began to shrink. "Snow-cap exceedingly small," he jotted in his logbook in mid-August. Then two months later, when it was summer in the south of Mars, the polar cap vanished. "No snow; certainly to speak of." Lowell believed, as did many in his time, that the white patches he saw on Mars consisted of frozen water, and he tried to suss out where the presumed moisture went when it melted. "So much water suddenly produced has got to be disposed of," he noted quizzically as the snows disappeared. Meanwhile, other features emerged. "Suspicions of canals," he wrote tentatively in his logbook.

Mars divulged its secrets reluctantly. To see the ghostly lines that Schiaparelli had traced on his maps took practice and patience. Even through the Lowell Observatory's powerful lens, even under cloudless skies, fine details on the Martian surface were difficult to discern because the image through the glass tended to wobble and blur, a problem due not to the telescope but to Earth's atmosphere. In space, the stars do not twinkle; celestial objects appear to shimmer because we view the heavens from the bottom of an ocean of air. Warm and cold currents in the atmosphere bend the light, as when summer heat rising from pavement seems to jiggle the horizon. The effect through a telescope is, as is everything, magnified. "To look for the canals with a large instrument in poor air is like trying to read a page of fine print kept dancing before one's eyes," Lowell remarked.

He stared at Mars for hours on end, waiting for rare moments when the image stabilized. During these flashes of clarity—mere fractions of a second—he glimpsed faint threads that stretched across the bright surface. At first, he doubted his eyes. "These sudden revelation peeps may or may not be truth," he inscribed in his logbook. Yet the visions returned, still tantalizingly brief. "Not sufficient time to make sure of suspected canals." Eventually, Lowell convinced himself that what he saw was real. "The number of canals increases encouragingly," he wrote his mother in September.

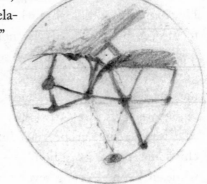

MARS. NOV. 4, 1894. (P. LOWELL)

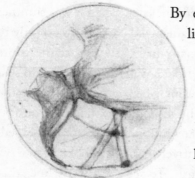

MARS. NOV. 19, 1894. (P. LOWELL)

By early November, he was seeing the lines "in profusion," and his sketches now depicted Mars as though a gargantuan spider had spun its silk upon the surface. "It looks as if it had been cobwebbed all over." Then, one late autumn's early twilight, on an evening so still that the smoke from Flagstaff's chimneys rose straight upward and a hush fell over the land, Lowell pointed his telescope at Mars and nearly gasped. Where he had previously seen single lines etched upon the planet, he now saw them twinned, like railway tracks across the western plains. Lowell fired a telegram to the press: "The canals of Mars have begun to double."

...

"NOW YOU SEE THAT the more one observes Mars the more the number of strange and inexplicable matters increases," the Italian who had first mapped the canals wrote a fellow astronomer around this time. Schiaparelli remained optimistic that scientists would eventually solve the mystery of Mars, but it seemed unlikely that he would be the one to do so. That summer, the Brooklyn science journalist Garrett P. Serviss, having already traveled to the moon with American theater audiences, was touring Europe to collect material for future lectures. He stopped in Milan, and when he arrived at the Brera Observatory to meet Schiaparelli, Serviss noted that this eminent scientist who had seen so far had now grown so nearsighted that he had to hold his guest's calling card five inches from his face. The colorblind astronomer was gradually going blind. It would be left to others to pick up Schiaparelli's mantle.

Elsewhere in Europe, the fabric of society continued to fray. Anarchists detonated more bombs—in Paris, London, and Barcelona—while in Lyon, a young revolutionary rushed the carriage of French president Sadi Carnot and plunged a dagger into his abdomen. (The fatal attack was revenge for the guillotining of the assassin's comrades.) In the United States, the lingering depression of 1893 convulsed the country with labor unrest. Populist businessman Jacob Coxey marched a ragtag band of the unemployed from Ohio to Washington, D.C., to demand federal jobs, his "tramp army" a seeming harbinger of class war. The shrewd union leader Eugene V. Debs, later a perennial Socialist Party candidate for president, then organized a massive rail strike that stranded passengers and freight, choking the very lifeblood of civilization until armed troops quelled the uprising. It all seemed further evidence of society's unspooling. "Our own little planet is proving itself, in these latter years of the nineteenth century, capable of furnishing to its inhabitants quite as much excitement as is good for them, without any assistance from Mars," a Rochester, New York, paper quipped. "Now what with . . . militant anarchy, Coxeyism, Debsism, and other portents, weak-minded people who are always on the look-out for the end of the world are under quite sufficient strain already."

Yet, in all seriousness, a connection existed between the science of astronomy and the public's end-of-the-century despair. Only a few generations earlier, it had been easy to believe that Earth was a hallowed world at the center of the universe with a caretaker God overseeing it all. Scientists had now shown that ours was an unexceptional planet around an ordinary sun, merely one star out of millions, and that objective laws and processes explained the workings of nature. People felt unmoored—alienated—and grasped at existential questions: the meaning of life, what happens when we die. Without God, who could provide the answers?

"Nothing, in truth, so much exalts our sense-perception, and at the same time admonishes and humiliates it, as the manifestations of astronomy," wrote the British author Sir Edwin Arnold. "We have enlarged enormously our conceptions of the universe, but apparently forgotten to magnify our beliefs." Arnold contended that where scientists had destroyed the old tenets of religion, it was now up to them to help build a new spiritual framework. "The best thing that could happen for mankind would be if a great astronomer had been born a poet or a great poet should become an astronomer," he concluded, "for we sadly need newer and nobler ideas in the chief of sciences."

Percival Lowell had always loved language, and, like many in his social class, he dabbled in poetry. As a child, when his toy boat sank in a puddle of snowmelt, he fashioned an ode to the shipwrecked vessel in Latin verse. In Japan, he wrote poems praising the beauty of cherry blossoms and the majesty of Mount Fuji, although his expatriate friends advised that he stick to prose. Now in Flagstaff, studying Mars, he opened a notebook and began to compose:

> A sister planet, whose sister face
> Complete in all its rounded grace
> Mirrors what we our Earth might see
> Could we once above it rise
> To behold it in its entity
> Sailing along through the pathless skies.

...

ATOP MARS HILL IN Flagstaff today, where Lowell made his original study of the planet in 1894, you can still find the Lowell Observatory—although vastly changed. What was originally intended as a temporary facility for a short-term scientific expedition has endured, growing into a thriving institution of research and education. I first made my way there in the summer of 2018, when Mars sailed exceptionally close to Earth, as I wanted to inspect the planet's surface through Lowell's very telescope, to see if I, too, might view the canals.

I had arranged for a private observing session beginning at 10:30 P.M. As I parked my car on pavement in the pines, I spied Mars through the branches—a speck of amber in the dark—then walked to the dome that houses the telescope. The current structure was built in 1896, and its interior looks like that of a barn, with wooden walls and a ceiling supported by exposed rafters and joists. Through the opening in the roof, I could hear insects and the night breeze.

The telescope inside resembled an enormous piece of artillery, like a cannon on a battleship, all gunmetal gray. An assistant placed a rolling stepladder by the great instrument's eyepiece. I ascended, set my right eye to the scope, and saw Mars bloom into brilliance like a full moon the color of apricot, or perhaps the sun near sunset. The planet throbbed and trembled—blurring, then jiggling, then occasionally holding still and coming into focus for a tantalizing instant. Watching and waiting for those moments of sharpness took intense concentration, and I tried to conceive what it was like for Lowell, hours spent at the telescope for mere glimpses of clarity while he was serenaded by the rhythmic chorus of the crickets, the soothing rush of the wind, the occasional wail of a train. What goes through one's mind in a setting like this? It seemed ideal for the inducement of dreamlike visions.

On the night I observed Mars, the fortunes were not on my side. I could see no polar cap, no dark patches, no canals because the planet was in the grips of a global dust storm, a veil that occasionally hides the surface features. I had better luck making discoveries during the day, in the observatory's archives. It was there that I came across Lowell's ode to Mars, tucked in a folder that had been stashed in a document case.

That poem from Lowell's first year in Flagstaff is a telling artifact. Sprawling across the pages of two composition books and demonstrating much rewriting—lines crossed out, words replaced—it reveals the man's obsessiveness. Lowell had grown bewitched by Mars. In 268 lines of meter and rhyme, he recast his scientific observations in heroic language. He pondered aloud what he had seen on Mars: the sparkling snows, the shrinking polar cap, the growing canals. His creative mind called on everything it had read and known: Schiaparelli's discoveries; Flammarion's flights of fancy; scientific notions of the formation of the solar system; Darwin's theory of evolution and its survival-of-the-fittest implications. The swirl of ideas was coalescing into a final question about life that may once have emerged on our neighboring world: "Are these very survivors dead? Or may Mars still be inhabited?"

CHAPTER
5

Lost World

1895

The city of Boston, long a hub of American culture, prided itself as a center of learning and literacy, so it was no surprise when, in the winter of 1895, massive crowds turned up for the opening of the latest edifice in Copley Square. At this crossroads one could find the Museum of Fine Arts, MIT, and Trinity Church, a roughhewn stone masterpiece by H. H. Richardson that inspired a movement in American architecture. Now visitors by the thousands came to admire a new and majestic structure, the Boston Public Library. Touted as a Beaux Arts "palace for the people," it drew gawking throngs who surged through the doors and up the marble staircase to gaze reverently at the vaulted rooms adorned with sculpture and paintings.

In this era still dominated by Queen Victoria and British imperialism, many Americans remained infatuated with England's legendary past—the medieval tales of King Arthur graced poetry and prose, drama and art—and one of the most celebrated elements of Boston's new library would be a monumental mural titled *The Quest and Achievement of the Holy Grail*. Its panels heroically depicted the Knights of the Round Table, those exemplary warriors whose ranks included a boy raised in the forest, naïve to the ways of men, who grew into chivalrous adulthood. His name was Sir Percival, and in some versions of the legend, it was he who found the Grail.

Percival Lowell, much like his mythical namesake, had gone on his own quest—an interplanetary one—and, having returned home, he was ready to relate his adventures and achievements.

Two blocks down from the Boston Public Library, in the dark and chill of a late-January evening, a stream of pedestrians made its way to MIT and entered the same auditorium where, a decade earlier, Edward S. Morse had lectured about the ways of Japan. The people now came to hear of an even more exotic destination. Ticketholders quickly found their seats, filling the rows that curved gently before the dais. Promptly at seven forty-five, ushers sealed the doors and the room quieted. Onto the stage of the Lowell Institute—an enterprise that was run, no less, by his own father—stepped Boston's poet astronomer.

> **LOWELL INSTITUTE.**
> PERCIVAL LOWELL, ESQ.,
> Will give four Lectures on
> **THE PLANET MARS:**
> RESULTS OF RECENT OBSERVATIONS.
> (With stereopticon views.)
> 1. Atmosphere. 2. The Water Problem. 3. Canals. 4. Oases, and the Doubling of the Canals.
> On Wednesday and Saturday evenings, at 7.45 P.M., beginning Jan. 23rd, 1895.
> AUGUSTUS LOWELL, Trustee.
> Tickets for these lectures will be ready for delivery at the Cadets' Armory, 130 Columbus Ave., on **Monday,** Jan. 21st, at 10 o'clock A.M.
> ja14-7t B. E. COTTING, Curator.

"Amid the seemingly countless stars that on a clear night spangle the vast dome overhead, there appeared last autumn to be a newcomer, a very large and ruddy one, that rose at sunset through the haze about the eastern horizon," Percival Lowell began in typically highfalutin prose. "That star was the planet Mars, so conspicuous when in such position as often to be taken for a portent."

No longer the dry speaker he had been as a graduating college senior, Lowell had matured into an orator who could command a room. His upright posture, his neat mustache, his patrician bearing conveyed a calm sophistication. He enunciated with precision.

Lowell explained that the proximity of Mars to Earth made the planet of special interest to scientists, for astronomers could study its surface in detail. "From [Mars], therefore, of all the heavenly bodies, may we expect first to learn . . . something in answer to the

MR. PERCIVAL LOWELL.

mute query that man instinctively makes as he gazes at the stars, whether there be life in worlds other than his own." In the course of four lectures over two weeks, he would attempt an answer.

Lowell opened with a basic question: Was it reasonable to conceive that life might exist on Mars? In other words, was the planet—if not inhabited—at least habitable? "Now, to all forms of life of which we have any conception," he asserted, "two things in nature are vital, air and water." In his first lecture, he demonstrated that Mars clearly had air, though the atmosphere was thin—far thinner than Earth's, even in the high Himalayas. "That beings constituted physically as we are would find it a most uncomfortable habitat is pretty certain," he acknowledged. "But lungs are not wedded to logic, and there is nothing in the world or beyond it to prevent, so far as we know, a being with gills, for example, from being a most superior person." Like the Frenchman who inspired him, Lowell seduced his audience with charm and wit. "To argue that life of an order as high as our own, or higher, is impossible, because of less air to breathe than that to which we are locally accustomed, is, as Flammarion happily expresses it, to argue, not as a philosopher, but as a fish."

A small item in the next morning's *Boston Herald* noted tepidly, "The lecture proved very interesting."

...

THREE NIGHTS LATER, LOWELL returned for his second lecture and resumed as if he had paused merely for breath. "After air, water. If Mars be capable of supporting life, there must be water upon his surface." Here he proceeded in a surprising manner. The public had long been told that Mars possessed water in abundance, for the planet displayed broad, dark patches that astronomers had labeled seas. "Several important facts conspire to throw grave doubt, and worse, upon their aquatic character," Lowell said. For one thing, the "seas" did not reflect sunlight the way water should. For another, the theory of planetary evolution, as Lowell saw it, suggested that Mars should be dreadfully dry.

As he had described long ago at his Harvard commencement, our solar system condensed out of a swirling cloud of gas and dust, and the planets began their lives as scalding, molten spheres. The smaller, lighter planets presumably lost their heat first, in the same way a teacup cools more quickly than a cauldron. Mars, therefore, matured into rocky adolescence—that is, it developed a solid surface and habitable climate—even while our own, much larger orb was still a burbling infant. "[Mars] is relatively more advanced in his evolutionary career," Lowell explained. "He is older in age, if not in years."

By this same reasoning, now that Earth had entered its evolutionary prime and was able to support life, Mars was presumably teetering toward infirmity. "As a planet grows old, its oceans, in all probability, dry up, the water retreating through cracks and caverns into the interior," he asserted. "Now, if a planet were at any stage of its career able to support life, it is probable that a diminishing water supply would be the beginning of the end of that life, for the air would outlast the available water. Those of its inhabitants who had succeeded in surviving would find themselves at last face to face with the relentlessness of fate—a scarcity of water constantly growing greater, till at last they would all die of thirst."

Was Mars, therefore, no longer habitable? Not necessarily. Given that the planet was evolutionarily "older" than Earth, Lowell suggested that life on Mars may have had a head start. Perhaps intelligence, indeed civilization, had emerged there eons ago, in ample time to adapt to the looming water crisis. Therefore, he asserted, if the planet possessed inhabitants, they had but one option to survive: "Irrigation, and upon as vast a scale as possible, must be the all-engrossing Martian pursuit."

Before concluding this second lecture, Lowell teased his audience with a brief turn from astronomical theory to observation. "At this point in our inquiry... the telescope presents us with perhaps the most startling discovery of modern times—the so-called canals of Mars. These strange phenomena, together with the inferences to be drawn from them, will form the subject of the next paper." After the crowd dispersed into the night, the following morning's Boston *Sunday Herald* offered a summary of what had been said but again expressed only marginal enthusiasm.

...

POPULAR NOVELS OF THE turn of the century, many of them—those written by men—told virile stories of high adventure at Earth's remote outposts. These books, termed *romances*, spun fantastic journeys through ancient, perilous, distant lands. Rudyard Kipling and Joseph Conrad were masters of the craft, as was Robert Louis Stevenson (one of Percival Lowell's favorite authors), whose *Treasure Island* filled the dreams of millions of boys and girls with peg-legged pirates and buried gold. The tale also sparked the imagination of novelist H. Rider Haggard, whose wildly successful *King Solomon's Mines* would itself inspire a whole literary subgenre about "lost worlds"—magical realms isolated by geography and time. Haggard set his stories in the uncharted core of Africa but feared that the mystery of that exotic continent would soon vanish. "Where will the romance writers of future generations find a safe and secret place, unknown to the pestilent accuracy of the geographer, in which to lay their plots?" he wondered. "Then the poor story teller... must betake himself to the planets."

For a public enamored of lost worlds, what Lowell showed at his third lecture must have seemed right out of fiction. Standing before his

THE PLANET MARS.

audience once again, he threw upon the screen a series of lantern slides that showed stylized drawings of Mars as he perceived it through his telescope. The orb appeared truly alien, a billiard ball crisscrossed by arrow-straight lines that met and diverged at common hubs. It looked as if the globe were networked by telephone wires or a railway system. These were Schiaparelli's mysterious canals, and Lowell believed he had unlocked their true nature.

If Mars really was a desert planet, if the only reliable water sat frozen at the poles, then bringing that moisture to temperate climes would be critical to the survival of the inhabitants. That is exactly what he believed the canals represented: a global irrigation system.

"At times the canals are invisible," he said. The lines tended to materialize during the Martian spring and summer, after the polar snows began to thaw. "Not until such melting has progressed pretty far do any of the canals, it would seem, become perceptible." Lowell emphasized that the lines themselves could not actually represent water, for the trenches would have to be of enormous width to be seen from Earth—fifteen miles across at a minimum. What astronomers called "canals," he claimed, were really broad strips of vegetation that greened up as the water flowed. Lowell had himself seen this phenomenon on Earth, in Egypt, where the Sahara bloomed on the Nile's banks.

Lowell then pointed, in his fourth and final lecture, to what appeared on Mars where the canals intersected: dark spots, most of them perfect circles. Other astronomers had referred to these dots as "lakes," but, like the canals, they darkened in the spring, just when plant life would green up, and they vanished in the fall, when vegetation would decay. "A solution of their character suggests itself at once," he concluded, "to wit, that they are oases in the midst of that desert." These were the settlements where the Martians tended crops to survive in the hostile, arid environment. It all amounted to one grand system. "The canals are constructed for the express purpose of fertilizing the oases," he said. What Lowell claimed to see on Mars was evidence of an ancient and superior culture, one higher on the ladder of civilization than any on Earth. "A mind of no mean order would seem to have presided over the system we see—a mind certainly of considerably more comprehensiveness than that which presides over the various departments of our own public works," he asserted.

Lowell made his case with such casual confidence and seeming logic that, had I been in that MIT auditorium in 1895, I can imagine myself being swept along. After all, those lines on Mars were an unsolved mystery, and Lowell's hypothesis offered a comprehensive explanation. Moreover, at this stage of his career as an astronomer, Lowell exhibited a whiff of disarming humility.

"All this, of course, may be a set of coincidences, signifying nothing," he conceded as he reviewed his evidence. He also acknowledged that many in his audience might reject his theory—although, he argued, not because it lacked logic but because it challenged the human ego. "Like the savage who fears nothing so much as a strange man, like Crusoe who grows pale at the sight of footprints not his own, the civilized thinker instinctively turns from the thought of mind other than the one he himself knows. To admit into his conception of the cosmos other finite minds as factors has in it something of the weird."

By granting that his ideas might seem fantastical, Lowell moved a step toward their acceptance. He softened the public's defenses.

...

BY THE TIME LOWELL concluded his series of talks in early February 1895, the newspapers had begun paying closer attention to what he had said. "It is impossible in print to describe the charm of Mr. Lowell's lectures," gushed Edward Everett Hale, venerable editor of the *Boston Commonwealth*. "His humor, his ready wit, his complete knowledge of the subject with which he deals, are such as one has no right to expect in the same public speaker."

Other journalists described the lecture course as nothing less than an adventure novel come to life. "It is as fascinating as a romance," commented Lilian Whiting, a prolific and influential Boston writer who penned weekly columns for western newspapers, in Illinois and Louisiana. Her readers comprised farm wives and progressive city women, including members of the New Orleans Lilian Whiting Club, who savored descriptions of the cultured New England life.

LILIAN WHITING.

"There's science and psychology, there's music, art, the drama, politics," Whiting once said, rattling off her long list of interests. She reviewed novels, took in the symphony, attended social clubs and all manner of theatrical performances. She had been among those at the Tremont Theatre in 1892 who traveled to the moon with Garrett P. Serviss. She now arranged an exclusive interview about Mars.

"What data have you deduced from your observations?" she asked Lowell.

"First, there is strong inference that Mars is a very living world, subject to an annual cycle of surface growth, activity, and decay; and second," Lowell continued, "this Martian yearly round of life must differ in certain interesting particulars from that which forms our terrestrial experience.... For unlike the Earth, which has water to spare, Mars is apparently in straits for the article, and has to draw on its polar reservoir for its annual supply."

Whiting marveled at the implications of an advanced civilization living next door. "What could more impressively lead to religious contemplation, to a new vista of reverence for the God of the universe, than that valuable series of lectures on Mars which Mr. Percival Lowell has recently concluded before the Lowell Institute?"

...

NEWS OF THE AUDACIOUS claims made by Lowell on a Boston stage seeped out into the broader country, slowly at first. Then came a surge.

"My lectures are coming out in the May, June, July and August *Atlantic*s," Lowell announced. Even before *The Atlantic Monthly* published its May issue, the magazine had begun promoting the series so heavily that when the issue hit newsstands, "the entire first edition was exhausted in a few days," Lilian Whiting noted. Although *The Atlantic*'s readership was small, its reach was massive. Newspapers across the country carried excerpts of Lowell's articles, distilled their essence, commented on their conclusions. "The rather wild speculations of Flammarion, based upon the observations of Schiaparelli, regarding the condition of the planet Mars, have been succeeded by rational and patient investigation," one paper prefaced its remarks. Millions read of Lowell's lectures, and as they did so, his ideas—his very language—filtered deep into the American consciousness. From New Jersey to California, Mars emerged as a topic for commencement addresses that June. "Is there anything to prevent an organism being created which would be adapted to [the Martian] thin air?" expounded one Helen Eldridge, a graduating senior at a small high school in upstate New York. "Even a person with gills might be a very superior person."

More than a few legitimate scientists were unamused by Lowell's claims. They found his reasoning thin, his conclusions hasty and absurd. A British astronomer estimated that digging the canals would have required "an army of 200 millions of men working for 1,000 of our years," an improbable feat even for superhuman beings. Edward Holden of the Lick Observatory accused his upstart competitor of falling victim to personal bias: "It is a point to be noted that the conclusions reached by Mr. Lowell at the end of his work, agree remarkably with the facts he set out to prove before his observatory was established at all." An astronomer in New York complained publicly that Lowell's ideas "cheapen the science of astronomy." Another, in Chicago, would privately ridicule "the good Sir Percival" who "with unsparing hand . . . sprinkles oases upon the barren wastes of Mars."

Presented with a choice between the stodgy old guard and an inventive newcomer, the public favored the latter. "There is no danger that astronomy will be 'cheapened' by the freest discussion of its problems," the New York *Sun*, one of the nation's most trusted papers, editorialized in Lowell's defense. "No great advance will ever be made unless the imagination is enlisted in the work." The Portland *Oregonian* agreed: "Though this article [in *The Atlantic*] reads like a fairy story, it proceeds by hypothesis as purely scientific and reasonable.... It constitutes what men of science call a working hypothesis." Garrett P. Serviss, that respected arbiter of science news, remained agnostic on whether Mars was inhabited, but he applauded Lowell. "[His] theory accounts very well for what is seen. At any rate, it is the most complete theory that has yet been advanced."

...

LOWELL DID MORE THAN advance a theory, however. He artfully spun a tale. Although the individual characters in this drama were too small to be seen through a telescope, one could imagine the pathos and heroism of this great civilization fighting to survive on a dying planet. The news-reading public at the turn of the century, beset by reports of anarchists and labor unrest and societal decay, could find much to admire in the Martians—these "men or creatures possessing the attributes of men," as one article described them, who "strive to ward off the annihilation which they know must in a little while overtake them." "What a colossal picture of mind at war with destiny!" exclaimed the British literary critic Andrew Lang. A prominent astronomy writer, Agnes Clerke, saw Lowell's theory tugging on humanity's heartstrings. "At the close of this nineteenth century, after so many poignant disillusions, amid the wreck of so many passionate hopes," she wrote, "men cherish the vision of other and better worlds, where intelligence, untrammelled by moral disabilities, may have risen to unimaginable heights."

Of course, Lowell's circumstantial case had yet to be proven, but a story's hold on the human mind has little to do with its truth or falsity. The legend of King Arthur endured in the popular imagination not because wizards, magical swords, and a place called Camelot really existed but because the Victorian public *wanted* them to exist. The ancient stories spoke to universal aspirations—bravery, honesty, loyalty, piety. In a similar way, the heroic story of the lost world of Mars was now burrowing into the larger American culture, imprinting itself on the collective psyche.

The Earth, meanwhile, spun upon its axis and flew in its orbit, bending summer to fall. In Boston, on stage at the Tremont, a modern interpretation of the King Arthur legend drew houses filled with the city's elite. In bookstores, just in time for the holidays, there appeared a handsome volume—"sumptuously bound in scarlet and gold," as Lilian Whiting put it—adorned with a simple, one-word title: *Mars*. It was Lowell's latest, comprising an expanded text of his Lowell Institute lectures, and it arrived to favorable reviews. "On the rostrum or the page, whether you agree with him or whether you do not, Mr. Lowell is always interesting," *The Brooklyn Citizen* opined. Between the book's covers one could find multiple drawings of Mars, and a map, which further enthralled readers. "I have studied [them] with an intense desire to make them as intelligible to me as those the old geographers made of Central Africa," the essayist Charles Dudley Warner wrote in *Harper's New Monthly Magazine.*

Lowell's book soon found its way into the collection of the Boston Public Library, down the street from where the words had been spoken the previous winter, but by then the author himself had left for Europe, hailed by *The Boston Sunday Globe* as a departing hero. "He is a scholar by instinct, and an astronomer by choice. He is rich and a bachelor, and he spends money without stint in carrying out his chosen work."

Lowell was soon, however, to navigate treacherous waters, both metaphorical and literal. In the early hours of Thursday, December 19, 1895, as his steamship approached its destination of Southampton, it veered to starboard when it should have turned to port and a grating noise from the keel announced that the vessel had lodged fast on a reef. "Fault of the pilot," Lowell grumbled, "aged 73 and bordering on imbecility." Lowell was soon carried safely to port by tug, but the incident lingered as an omen. Within a few days, he would arrive in Paris and receive a warning meant to guide his own reputation away from rocky shoals.

CHAPTER
6

Allies and Adversaries

1896–1897

Camille Flammarion spent the warm months of each year just south of Paris, at his private estate in Juvisy, but winters found him in the city, in a sunlit flat atop a five-story apartment block at 16 Rue Cassini. The street had been named for the Italian astronomer Giovanni Domenico Cassini, whom King Louis XIV had lured to France in 1669 to help build and run the Paris Observatory. That institution, adorned with its great dome, loomed outside Flammarion's corner window. Inside the apartment, the décor left no doubt that this was home to a man whose mind embraced the cosmos. The dining room ceiling had been painted sky blue, with clouds. Below, leather chairs sported signs of the zodiac. The Greek muse of astronomy, Urania, smiled down

THE DRAWING-ROOM IN
M. FLAMMARION'S PARIS HOME.

from a decorative plate above the buffet, while in Flammarion's study shelves of books, heaps of scientific reports, and two small telescopes offered a sense of scholarly clutter. A cushion on the living room sofa expressed more discreetly the prodigious output of the French astronomer's remarkable head. Still lusciously maned at fifty-three, Flammarion was as prolific at growing hair as he was in his writing. "There is where I put his hair after cutting it," confessed his wife, Sylvie, gesturing at the overstuffed silk pillow.

One day toward the beginning of 1896, Flammarion was at home when he received word that a caller had arrived at the door. "Sir, it's an American astronomer who asks to speak with you." Flammarion instructed that his guest be shown in, then greeted the lanky sophisticate who stepped inside. "How delightful to have you, Mr. Lowell. We in France know your fine work in astronomy." The two men had corresponded over the previous year, but this was their first meeting, one that would initiate an enduring friendship, and it clearly struck Flammarion as newsworthy—enough so that he recounted the interchange in an article for a French weekly.

"Have you come to see Europe?" Flammarion asked.

"I'm on my way to the Sahara," Lowell replied. Having observed Mars from Flagstaff the last time the planet came close, Lowell was exploring the possibility of moving his observatory to North Africa—to a site with an even clearer, calmer atmosphere—in time for the planet's next approach, the coming December. "But first I want to show you what we've already accomplished in our Arizona mountains," Lowell continued. "It was your book on the planet Mars that gave us the impetus."

Lowell explained that he, too, had written a book about Mars, and he presented his host with an early copy. Tucked in the back was a folding map of the planet as observed from Flagstaff. Lowell pointed to some marked changes from previous charts drawn by Flammarion and Schiaparelli, notably the profusion of canals. "You have only 79 on your map," Lowell said. "We have 183. That's 104 new ones." Some of Lowell's canals crossed the dark areas—the Martian "seas"—showing

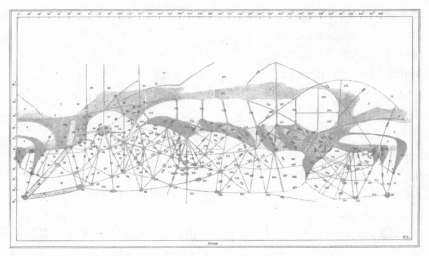

MAP OF MARS ON MERCATOR'S PROJECTION.

that they could not be water (as Flammarion had believed). And at the intersection of the canals sat Lowell's oases, the dark spots previously called "lakes" but which proved to be "almost all perfectly round." Lowell summarized his theory: that Mars was largely desert, that the canals were irrigation ditches lined by vegetation, that the oases were the Martian settlements.

"Look at the geometric design," he said. "It's intentional. This was made expressly to drain the water." Lowell pointed to a prominent oasis. "Doesn't that look like a major city? And here, too."

Flammarion needed no convincing that Mars might well be inhabited, but even he approached the new theory, and new map, with caution. The planet had a history of playing tricks on astronomers, including Cassini, the famous Italian whose name sat affixed to the street outside. In 1666, Cassini had carefully observed Mars and noted distinct shapes on its surface, then measured how long these features took to rotate out of view and back around into position. This enabled him to calculate the length of the Martian day: twenty-four hours and forty minutes, an estimate that proved remarkably accurate. Remarkably inaccurate, however, were Cassini's sketches. On one hemisphere of the planet he drew a prominent surface feature that looked like two round lobes connected by a fat line.

"You know as well as I do," Flammarion reminded Lowell, "that the dumbbell figure drawn two centuries ago by Cassini doesn't exist."

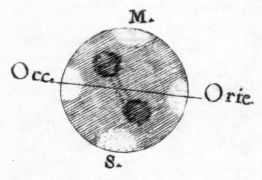

CASSINI'S SKETCH OF MARS.

It was a lesson to be heeded. "At the limit of visibility, we may be fooled by illusions, and even afterward continue to see what we thought we saw the first time." Flammarion felt compelled to ask: "Are you sure you saw correctly?"

"We are perfectly sure," Lowell answered with conviction. "And now we confidently submit our observations for other astronomers to review."

...

LIKE ANY HUMAN ENDEAVOR, science can be political, and fiercely so. Coalitions form. Animosities develop. Theories rise and fall as much on the reputation of the proponents as on the quality of the evidence. Scientists who hold similar notions tend to aggregate, forming self-selected groups of like-minded individuals who reinforce each other's shared opinions. In the case of Mars, this sorting into factions would eventually create two camps, later known as the *canalists* and the *anti-canalists*. Lowell would rapidly rise to lead the former group, which saw the canals not only as real features but as possible evidence of a Martian civilization. The Lick Observatory's Edward Holden was, for the moment, the most aggressive and outspoken opponent. War would ultimately break out between these groups, but for now many astronomers, particularly in the Old World, had yet to take sides. Lowell was on a campaign to woo allies.

Back in England after meeting with Flammarion, Lowell gladhanded Herbert Hall Turner, observatory director at the University of Oxford, to whom Lowell also delivered a copy of *Mars*. Turner found himself enthralled. "I will risk all the horrible suspicions of bribery and cordially advise others to get the book," he cheerfully urged his colleagues. "They will, I feel sure, soon feel the fascination of the story told."

Lowell also journeyed to Milan, to track down the man who had discovered the Martian canals. While Schiaparelli long believed in the reality of the lines on Mars, he often suggested that they might be natural features, not signs of intelligence. Lowell nudged him the other way. When Lowell displayed his new map and pointed to the Martian settlements—the oases—the Italian astronomer's failing eyes lit up. "I suspected them myself," he said, "but could never see them well enough to make sure."

Before long, Schiaparelli grew more imaginative and outspoken. One of the enduring Martian mysteries was the doubling of the canals, a phenomenon that even Lowell's theory could not explain. "Here, the intervention of intelligent thought seems well indicated," Schiaparelli wrote in an article he sent Flammarion for publication in France. As Schiaparelli saw it, the Martians might have constructed parallel irrigation furrows down long valleys, the paired ditches running at different elevations. In the Martian spring, the uppermost channels are filled with water, and the vegetation grows along these single lines. In the summer, when the flow from the melting polar ice intensifies, the "Minister of Agriculture orders the highest locks to be opened and fills the two upper canals with water[,] . . . the valley changes color in these two zones, and the astronomer on Earth perceives a gemination." Schiaparelli remained coy, however, about the seriousness of his suggestion. Before sending his article to Flammarion, he wrote atop the first page, in Latin, "*Semel in anno licet insanire.*" Once a year it is permissible to be mad.

. . .

WHILE LOWELL'S EFFORTS TO lobby European astronomers yielded some success, his work in North Africa proved a definitive failure. In Algeria, at the time ruled by France, he scouted for a possible new location for his observatory but abandoned the project when he found that the stars twinkled badly, revealing an unsteadiness in the atmosphere that would make Mars blur and wobble through a telescope. The view on the ground, however, was clearly inspiring. "Palms choke the outlook as I write," Lowell penned in a letter home, his fine handwriting gracing the elegant stationery of his upper-class hotel. Ignoring the violent history of French dominion here and the continued subjugation of the Algerian people, Lowell seemed giddy to find himself in an exotic outpost on the edge of a vast desert. "Arabs everywhere in picturesque squalor and beautiful bronze skins," he wrote in the romantic yet racist language of Victorian orientalism. "I feel as if I were vouchsafed half-visions of the Martians in their perpetually sun-lit planet and oasis-like life."

This conception of Mars as serenely Mediterranean was taking hold in other quarters too. "On arriving at one of these 'oases,' . . . what should we see?" Garrett P. Serviss asked rhetorically as he imagined a visit. "Around us a brilliant scene of vegetation, a blaze of floral colors, golden fields of nodding grain, luxuriant gardens filled to overflowing with rich, quick-growing fruits and musical with singing birds; on every side as far as the eye can extend, a scene of tropical splendor, and in the midst the gilded minarets of a city more beautiful than Damascus or Seville rising into a sky as blue as an unruffled sea." The Boston *Sunday Post* offered a similar description, its artist's

THE *POST* ARTIST GIVES A SUGGESTION OF WHAT THE PEOPLE MAY LOOK LIKE ON THAT HEAVENLY BODY.

illustration portraying the Martians as if ancient Romans—clad in togas, surrounded by palm trees, conversing with a look of wisdom on their faces.

...

UPON HIS RETURN FROM North Africa in March 1896, Lowell continued his vigorous work. He expanded his staff of trained assistants. He upgraded his observatory with a more powerful telescope. He temporarily moved the telescope to Mexico in search of clearer skies—then back to Arizona—attracting a whole new band of admirers. "Mars ought to appreciate the devotion of Prof. Lowell," a Mexico City paper opined. "She surely could not have a more faithful or chivalric knight."

Meanwhile, Lowell grew more self-assured in his pronouncements. "I have no doubt that there is life and intelligence on Mars," he said on his way from Boston to Arizona, itself a scene of violent struggle for existence. Lowell arrived in Flagstaff to news from across the territory—chronicled in the local paper—of a train robbery, a murder, even the dynamiting of a saloon, yet his own focus remained heavenward as he once again watched the canals emerge in the Martian spring and double in the Martian summer. "The significant point of the whole is that the seasonal phenomena of the Martian year are repeating themselves at this present one, even to details," he telegraphed *The New York Herald*. "That Mars is not only a very living world, but a very orderly one to boot, is pretty conclusively shown by the phenomena."

Despite the outward displays of confidence, Lowell was on the verge of another psychological break, much bigger than his first. What provoked his collapse is not entirely clear. One factor may have been the recent death of his mother, his greatest champion and consoler. Another may have been the pressure he undoubtedly felt to prove his grandiose claims, especially given his father's exhortation from childhood to achieve something "of real significance." Then there was the matter of Lowell's own hardwiring, for his frequent bouts of manic activity suggest he may have lived with bipolar disorder. The most likely trigger for his collapse, however, was the blistering reception Lowell received when he turned his telescope from Mars toward Venus.

Just as Cassini in the seventeenth century had measured the length of the Martian day by tracking surface features that rotated in and out of view, Lowell attempted to do the same for Venus, Earth's other close neighbor. He studied the planet and carefully drew what he saw: a network of lines radiating like spokes from a central point. The sketch struck many as absurd, for it was believed, and is true, that the surface of Venus cannot be seen from Earth; it is cloaked in a thick layer of clouds.

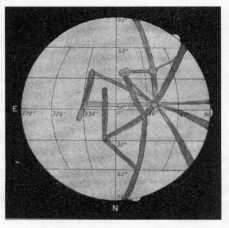

CHART OF VISIBLE HEMISPHERE OF VENUS. (P. LOWELL)

"This looks to me suspiciously like Mars," a veteran British astronomer scoffed when he saw that the neophyte from America had drawn lines like the canals. "I do not know whether Mr. Lowell has been looking at Mars until he has got Mars on the brain." Other critiques predictably issued from California's Mount Hamilton, delivered by the Lick Observatory's gruff director, Edward Holden. "I have no hesitation in saying that such markings as are shown by Mr. Lowell [do] not exist on Venus," Holden fired with typical contempt. "They are illusions." More surprising, and presumably more cutting to Lowell, were the reviews that emanated from Flammarion's observatory in Juvisy. "We cannot help considering the whole of this anomalous canal system [on Venus] as entirely illusory," the French astronomer's chief assistant asserted.

Lowell had failed to heed Flammarion's cautionary tale of Cassini's dumbbell on Mars; he had put too much faith in his all-too-fallible eyes. Indeed, what Lowell saw on Venus may well have been the blood vessels of his own retina, which, given the settings of his telescope, may have looked like enormous lines on the distant orb. Lowell would eventually acknowledge his error (although, even later, he would retract the retraction). His was a devastating mistake, one that raised questions about his judgment, his eyesight, and the reliability of his maps, including those of Mars.

Lowell slumped into melancholy. "I have had what I never supposed would overtake me, a complete breaking down of the machine," he wrote from Boston in the spring of 1897. By late summer, he tried to return to Flagstaff but failed to muster the energy. "Am pretty weak," he confessed. "Case of evil-ution."

Usually a man-about-town, Lowell vanished from the society pages: no polo matches, no nights at the theater, no Brahmin weddings. Struggling with what today would likely be diagnosed as depression, Lowell had come down with a condition widely known then as *neurasthenia*, whose sufferers complained of sleeplessness, hopelessness, crippling anxiety, and an inability to focus. The ailment was said to result from a depletion of one's stores of nervous energy, as if the patient had literally drained the voltaic battery that kept the body running. It often struck individuals of intellect and accomplishment and was frequently blamed on the fast pace of late-nineteenth-century life. Treatment ranged from electric shocks to powerful tonics made with arsenic and strychnine, but the first and most important prescription was rest.

Doctors placed Lowell in the care of his father and forced him into bed for a month—no visitors, no work—which Lowell contended only made his condition worse. He next sailed for Bermuda, as neurasthenics were often said to benefit from sea air and a change of scenery. Lounging at the Hamilton Hotel, his recovery continued to drag, though he tried to sound upbeat in a letter home, characteristically peppering his English with Latin phrases: "*Festina lente* [make haste slowly] is nature's motto for me, and I try to make *nulla vestigia retrorsum* [no steps backward]." When Schiaparelli learned that his American friend was ill, he wrote encouragingly from Milan, "Everyone truly hopes that [your health] will improve so that you may be able to continue the work you've begun in such a brilliant and grandiose

HAMILTON HOTEL.

manner." It was unclear, however, when or if Lowell would return to the scientific battlefield.

Perhaps the only consolation for Lowell was this: his pugnacious adversary was hardly faring much better. Edward Holden, the Lick's dictatorial director, had by now so alienated his own assistants that they conspired to, as one mutineer colorfully put it, "rid the Observatory of its incubus." A San Francisco newspaper announced the outcome in the fall of 1897—DIRECTOR HOLDEN RESIGNS—and the once-mighty astronomer slunk away, unemployed, his reputation sullied, the doors to other scientific opportunities closed before him.

The leading warriors of the amassing armies were now sidelined. Further combat over Mars would have to wait. Other battles, however, both real and imagined, were about to begin.

CHAPTER

7

War Stories

1897–1898

With Percival Lowell paralyzed by neurasthenia, another player now emerged to sear the Martians into cultural consciousness. He was an Englishman, a young writer who had recently penned an anonymous essay in London's *Saturday Review* arguing against the notion that the hypothetical people of Mars were just that: people. "No phase of anthropomorphism is more naïve than the supposition of men on Mars," he had insisted. "There is every reason to think that the creatures on Mars would be different from the creatures of earth, in form and function, in structure and in habit, different beyond the most bizarre imaginings of nightmare." The unidentified writer did not disclose that he had been pondering Martians and their potential form for a practical reason: they would play a central role in a book he was writing, a work of fiction that would forever tinge the night sky with terror.

Herbert George Wells—"Bertie" to his parents, "H. G." to his friends—was at the time living in suburban London, in the county of Surrey, in a town called Woking. In his late twenties and on the cusp of meteoric professional success, he had moved from the city the previous year to escape its bad air and to flee the scent of scandal that arose when he left his wife (whom he soon divorced) for his mistress (whom he soon wed). Wells hoped that this more bucolic setting would prove calming and conducive to work. He took up residence in

H. G. WELLS.

a two-story semidetached home across from the railway on the Maybury Road, and he wrote furiously, as if he were a man running out of time.

Wells was the literary converse of Lowell the scientist; where Lowell brought a poetic sensibility to his astronomy, Wells applied strict science to his writerly imagination. In college, he had studied under the great biologist Thomas Henry Huxley, remembered as "Darwin's bulldog" for his fierce advocacy of evolutionary doctrine. Wells was now working as a journalist and an aspiring novelist—churning out book reviews, theater reviews, short stories, and essays, and returning time and again to questions about life and natural selection. His mind was a crucible, testing and mixing ideas. In a piece titled "Concerning the Nose," he appraised the evolutionary significance of that "salient angle," the human proboscis. In "The Living Things That May Be," he mulled the possibility of life beyond Earth. ("Can there be anywhere in the infinite ocean of space other little islets of hope and pain?") In "Through a Microscope," he waxed philosophical while examining tiny life-forms under a lens. "And all the time these creatures are living their vigorous, fussy little lives in this drop of water they are being watched by a creature of whose presence they do not dream, who can wipe them all out of existence with a stroke of his thumb."

An atheist and socialist, Wells embodied the anguished end-of-century mindset. He viewed civilization as a thin veneer that could easily crumble to dust. One day his eldest brother, Frank, gave him an idea for a story: "Suppose some beings from another planet were to drop out of the sky suddenly and began laying about them here!" Wells remembered that remark on a later stroll through Woking, and an image came to mind, of the ground cratered by metal cylinders that had arrived from some other world. *What planet should I choose?* he thought. The answer was obvious. *Mars, of course.*

Earth was in the midst of a revolution in transport. Recent advances in the design and manufacturing of bicycles had suddenly made them a safe, affordable, and fashionable mode of travel. "They are on every street, they are in all the parks," Lilian Whiting wrote from Boston while, in Paris, Camille Flammarion was spied "spinning down" a boulevard "with his hands behind his back and an air of evidently keen enjoyment." The bicycle symbolized freedom and modernity (and even suggested a new explanation for Schiaparelli's mysterious lines on Mars; CANALS ARE REALLY BICYCLE PATHS one newspaper teased in a headline). H. G. Wells was also caught up in the cycling craze, and he took daily rides to clear the cobwebs from his mind. He pedaled the Surrey byways, through undulating terrain and honeysuckle-scented air, wending past frozen-in-time villages, fields of sheep, and homes of red brick. And as he cycled the pastoral scene, he gleefully pictured it incinerated by visitors from outer space. "I'm doing the dearest little serial," he wrote an old college friend, "in which I completely wreck and

destroy Woking—killing my neighbours in painful and eccentric ways."

Publication of the tale was slated to debut in *The Cosmopolitan* magazine in the United States, and *Pearson's* in Britain, the coming spring. An excited Wells alerted his brother Fred, who was in South Africa at the time. "I don't know if you see *Pearson's Magazine* out there—in April next a long story of mine will begin and go on until December," Wells wrote. "I expect great things for it." Indeed, *The War of the Worlds* would become a sensation.

...

FROM ITS VERY OPENING, the story snared readers with one of the most haunting passages in modern fiction, the words and ideas a natural extension of the fertile daydreams contained in Wells's recent essays:

> No one would have believed in the last years of the nineteenth century that human affairs were being watched keenly and closely by intelligences greater than man's and yet as mortal as his own; that as men busied themselves about their affairs they were scrutinized and studied perhaps almost as closely as a man with a microscope might scrutinize the transient creatures that swarm and multiply in a drop of water. . . . Yet, across the gulf of space, minds that are to our minds as ours are to the beasts that perish, intellects vast and cool and unsympathetic, regarded this earth with envious eyes, and slowly and surely drew up their plans against us.

Decades later, this masterwork would be famously retold on the radio by the American actor and theater producer Orson Welles, prompting some number of listeners to actually flee their homes, convinced that an alien attack was under way. In its original incarnation, however, *The War of the Worlds* caused no such panic. No one could have mistaken the story for news. The drama played out slowly, published in magazine installments that appeared monthly through the spring, summer, and fall of 1897.

"There is no more absorbing serial running than Mr. H. G. Wells' 'War of the Worlds,'" one enthusiastic reader remarked. Another, luridly drawn by its morbid aspects, seemed to agree: "The present installment is rather grewsome [sic], and makes one imagine the sort of nightmare which might result from a mixture of champagne supper and Darwinism." Readers on both sides of the Atlantic were held rapt. That November, one who had been following the tale with special interest—he worked in a seven-story building in downtown Duluth—heard a loud thud and feared for a moment that the Martians had arrived. It was a tub of butter that had fallen from a high window.

Wells's fictional account, though utterly fantastic, felt intensely real. The novelist had cleverly crafted the narrative in autobiographical form, telling the story from the perspective of a man living outside London, in the town of Woking, who had experienced the interplanetary battle firsthand. The invasion began with the incendiary fall of a shooting star. It turned out to be an enormous metal cylinder that had cratered the ground nearby, in an area of open heath and sand surrounded by pines. Crowds gathered out of curiosity, then were torched by heat ray as the Martians emerged. The aliens soon rose up in enormous tripods—three-legged killing machines that rampaged toward London, fumigating the streets with poisonous smoke while hunting human prey. Civilization devolved into savagery as millions stampeded in panic, until the tripods mysteriously broke down. The final Martian made its last stand on Primrose Hill, a grassy slope that looked across the great and ruined city. Here the dying creature issued a plaintive wail—*ulla, ulla, ulla, ulla*—then fell silent, conquered not by feeble humans but by mighty bacteria.

Wells deliberated carefully on what the Martians inside the machines should look like. He began with the concept, espoused by Lowell and other astronomers, that Mars was evolutionarily "older" than Earth. This implied that the red planet's inhabitants might resemble the earthlings of the far future. In other words, the Martians *are* what we humans *will be*, after evolution has shaped us many generations down the line.

Wells had already performed this thought experiment in an earlier piece called "The Man of the Year Million." He had postulated that in the coming millennia, as people grow more dependent on technology, natural selection will favor our minds over our bodies, eventually producing a creature with a large head, shriveled appendages, and no need for stomach or intestines (for surely our distant descendants will find a more advanced way to consume nutrients than by eating). This future human became the basis for his Martian: an asexual monster that looked like a cross between an octopus and a jellyfish, with sixteen whiplike tentacles. The animal was little more than a brain enveloped in a sac of flesh. It survived by injecting the blood of its prey into its veins.

THE ACTUAL MARTIANS WERE THE MOST EXTRAORDINARY CREATURES IT IS POSSIBLE TO CONCEIVE.

Horrifying, vivid, and at turns playful, Wells's story drew its power from being anchored in real geography and true science. "Had our instruments only permitted it we might have seen the gathering trouble far back in the nineteenth century," the unnamed narrator reflected, gazing toward the past from his future perch. "Men like Schiaparelli watched the red planet . . . but failed to interpret the fluctuating appearances of the markings they mapped so well. All that time the Martians must have been getting ready." The fictional storyteller concluded that humans had missed the ominous signs, so evident in retrospect. As he noted, astronomers had seen a light on the rim of Mars in 1894, much like the triangle observed during the boom of 1892. "I am inclined to think that that appearance may have been the casting of the huge gun, the vast pit sunk into their planet, from which their shots were fired at us."

. . .

DURING THE PERIOD THAT Wells's fictional war played out in the magazines in 1897, a real conflict brewed across the ocean. Spain, once a colonial behemoth as the dominant European power in the New World, now clutched at one of its last remaining possessions in the Americas, the island of Cuba. Rebels fought for independence while their Spanish overlords brutally corralled civilians into concentration camps, where the starving people withered to skeletons and perished in untold thousands. In the United States, just ninety miles away, President William McKinley urged peace, but his new assistant secretary of the navy, Theodore Roosevelt—in-law of Percival Lowell's widowed sister—lobbied for armed intervention, for the sake of the Cubans as well as American morale. "In strict confidence," Roosevelt wrote another military man, "I should welcome almost any war, for I think this country needs one."

Also beating the drums of war was the American press, especially the cunning young publisher William Randolph Hearst—a wealthy son of privilege who, like Lowell, had attended Harvard (although Harvard had expelled Hearst for his carousing and poor grades). Leveraging his mother's money and his own affinity for grandiosity, Hearst had recently bought the *New York Journal* and proceeded to outdo the sensationalistic journalism of Joseph Pulitzer's *World*. Hearst blithely and willfully manufactured news, enlarged headlines to outrageous proportions, and launched all manner of stunts in a full-throttle bid for readers, practices that the more conservative press derided as "yellow journalism." With his deep pockets, Hearst hired star talent—novelist Stephen Crane to report on New York prostitution, Mark Twain to cover Queen Victoria's Diamond Jubilee—and he raided Pulitzer's newsroom for its top journalists.

One of Hearst's recent trophies was Arthur Brisbane, a shrewd editor who under Pulitzer had participated in *The World*'s famously hyped Mars coverage in 1892. Now in charge of Hearst's *New York Evening Journal* and with his salary tied to its circulation, Brisbane devised creative schemes to boost readership, and he once again leveraged public fascination with the red planet. In December 1897, he republished *The War of the Worlds* as a newspaper feature, in bite-size portions that lured readers back day after day for a month. Although he presented the story as fiction—in other words, truthfully—he slyly, and without the knowledge or permission of H. G. Wells, changed the alien invasion's setting from London to metropolitan New York.

At a time when the American public feared an impending war with Spain, readers gasped at breathless descriptions of the Martian tripods laying waste to Manhattan: demolishing Columbia University, pulverizing the Brooklyn Bridge, toppling one of the lofty spires of St. Patrick's Cathedral. The gimmick drew an audience, which spurred

THE DESTRUCTION OF BROOKLYN BRIDGE.

another popular yellow journal, *The Boston Post*, to copy the prank in modified form, now turning the Martians' wrath on eastern Massachusetts. In this version of the story, the palatial Boston Public Library, so recently constructed, now "collapsed itself and buried beneath its ruins thousands of rare and valuable books and priceless works of art."

Not to be outdone, the *New York Evening Journal* then extended its Martian hijinks. The editor, Brisbane, asked an old colleague to craft a sequel to *The War of the Worlds*. The result was *Edison's Conquest of Mars*, a bit of patriotic fluff that proved ideal for Hearst's jingoistic paper (and was then reprinted in more than a dozen other dailies across the country). The author, branching out from journalism to fiction, was the imaginative Garrett P. Serviss, and while he was no literary artist, his story offered a satisfying postscript to Wells's alien invasion. "Let us go to Mars. We have the means," came the cry of vengeful humans in the first installment. "Let us ourselves turn conquerors and take possession of that detestable planet, and if necessary, destroy it in order to relieve the earth of this perpetual threat which now hangs over us like the sword of Damocles." Over the next month, readers followed the tale of Earth's retaliatory strike on its warmongering neighbor, a global effort led by a fictionalized Thomas Edison. A squadron of the inventor's antigravity airships, outfitted with disintegrator guns, arrived at Mars and eventually identified a weak point in the planet's defenses: the canal system. Edison destroyed a key hydraulic station, which triggered a planetwide flood that drowned the Martians and washed away their cities.

Just five days after the story concluded in the *Evening Journal*, the already tense situation in Cuba slouched toward outright war. An American battleship stationed in Havana, the USS *Maine*, exploded and sank, taking down with it more than 250 American lives. No one knew if Spain had set the deadly blast, but that conclusion was widely inferred—and was certainly implied by the yellow press—spurring the public to demand revenge. Even as experts warned that the U.S. military was ill prepared to confront the European power, America surged with patriotic fervor. "What's all this talk about Uncle Sam not having a navy big enough to fight Spain?" one of Hearst's readers wrote defiantly. "These fellows seem to have forgotten all about Mr. Edison and his air ship, to say nothing of those disintegrators that played havoc with Mars and the Martians."

Through the spring and summer of 1898, the United States battled Spain across the hemispheres in what would come to be known as the Spanish–American War (though more than one newspaper at the time called it "McKinley's War of the Worlds"). Adding to the international drama was Theodore Roosevelt's ragtag band of Rough Riders that made its much-ballyhooed charge up San Juan Hill. "Oh, but we have had a bully fight," Roosevelt declared to cheering crowds as he returned home a hero. A war that had begun for ostensibly humanitarian reasons—yet was propelled, as well, by strategic interests—transformed America from a democracy that had been born out of its own war of independence into an imperial power. The nation now controlled not only Cuba but also Puerto Rico, Guam, and the Philippines.

Also victorious when the fighting ended were William Randolph Hearst and his ilk, who had seen their newspaper readership soar even in the war's prelude. "As we see to-day, in spite of all the ridicule that has been lavished on the 'yellow journals,' . . . their circulation is apparently as large as ever," the dignified New York *Evening Post* noted with regret.

. . .

THE YELLOW PRESS HAD changed journalism with its mix of hype and sensation, spectacle and fiction, and it continued to focus on Mars. In late 1898, a journalist in Chicago tried to engage a local astronomer in a discussion of the red planet. "Do you think the canals upon Mars are artificial?" the reporter inquired of S. W. Burnham, a specialist in double stars.

"We don't know that there are canals upon Mars," Burnham answered dryly.

The newsman pressed on, undeterred. "What degree of dissimilarity to ourselves is it possible the inhabitants of Mars possess?"

"We don't know that there are inhabitants upon Mars," Burnham replied.

"Well, if there are inhabitants upon Mars," the journalist persisted, "is it probable that they have built canals that can be seen from this earth?"

"My dear sir," Burnham responded with evident disdain, comparing the reporter to a dimwitted aristocrat in the British stage farce *Our American Cousin*, "you recall to my mind one of Lord Dundreary's droll questions. 'Does your brother like cheese?' 'I have no brother.' 'But if you had a brother would he like cheese?'" At this, the astronomer put the conversation to rest. "Nothing, I repeat, is absolutely known about Mars' inhabitants, if it have any, nor about its canals, if it have them."

For the newspapers to continue to tell sensational stories about Mars, with Lowell still out of the public eye, the yellow press needed a new scientific partner—one willing to feed its appetite for mystery and wonder.

CHAPTER
8

WIRELESS

1899–1900

IF EVER A CHARACTER HAD BEEN TAILOR-MADE TO THE FASHion of the yellow press, it was Nikola Tesla. An inventor of strange brilliance, Tesla cultivated an air of mystery and refinement. At times flush with income from patents and the backing of millionaire investors (but at other times nearly broke), he lived in that hyphenated symbol of New York luxury, the Waldorf-Astoria, where he befriended the wealthy and the famous. Mark Twain and other celebrities frequented Tesla's Manhattan laboratory, a veritable sorcerer's den where what appeared to be magic became reality. Artificial lightning crackled in the dark. Glass bulbs glowed without any evident power source. Ghostly new emanations called X-rays revealed the inside of the human body: ribs, shoulders, heart, skull. Tesla willingly sacrificed most anything for his work—his money, his reputation, his health. He

NIKOLA TESLA.

engaged in self-experimentation, bathing his brain in radiation and jolting his body with electric shocks, the latter an attempt to ward off neurasthenia, the depressive condition that had disabled Percival Lowell. "You see," he explained, "electricity puts into the tired body just what it most needs—life force, nerve force."

Tesla exerted his own force upon the New York press, which eagerly published eye-popping stories from the realms of technology and the natural world. "Ain't science wonderful!" exclaimed Arthur Brisbane, the crudely ambitious editor at Hearst's *New York Evening Journal* who, besides commissioning *Edison's Conquest of Mars*, was an early booster of Tesla's. Anything loosely defined as "science"—the more bizarre, the better—proved ideal fodder for the papers. This was especially true for the thick Sunday editions that, in a bid to appeal to the whole family, overflowed with gee-whiz stories of sea monsters and dinosaurs, cannibals and wild beasts. Tesla's grandiose claims provided frequent material for sensationalistic headlines:

TESLA SAYS MEN MAY YET LIVE FOR CENTURIES.

TESLA'S ELECTRICAL SUBSTITUTE FOR THE BATH-TUB.

TESLA'S LATEST DISCOVERY:
FERTILIZING IMPOVERISHED LAND BY ELECTRICITY.

TESLA'S GREAT INVENTION TO BLOW UP HOSTILE WARSHIPS
BY ELECTRIC WAVES.

The more hidebound periodicals of the era protested. "Whole pages of the yellow journal seventh-day editions are loaded down with pseudo-scientific pabulum," *Scientific American* grumbled. "We are not alone in our expressions of regret that any one of Mr. Tesla's undoubted ability should . . . be flooding the press with rhetorical bombast." Tesla was a man of undeniable drive and accomplishment, so married to his laboratory that the press reported he had no time for romantic involvement with women. (Unremarked in the papers at the time was his affinity for men.) He had invented a revolutionary new motor and, together with George Westinghouse, transformed the use of electricity by demonstrating the superiority of an alternating-current scheme of power generation and distribution. Thanks in part to Tesla, electricity could be transmitted long distance by wire, making it far more available and affordable than with Edison's direct-current technology.

Tesla had now turned to a new infatuation, devising ways to transmit electric power and signals *without* wires. By the late 1890s, he had begun experimenting with what would later be known as radio but was then referred to as wireless telegraphy, for the goal was to transmit Morse code messages through the air. Prior to 1895, no such signal had been sent more than a trifling distance, but then—seemingly out of nowhere—a young inventor in Italy began breaking records. Guglielmo Marconi flung his electrical signals a hundred feet, a hundred yards, a mile and a half. Shifting his experiments to Britain, Marconi soon achieved even greater feats: eight miles, eighteen miles, then thirty-two miles when he astonishingly flashed wireless pulses across the English Channel.

Tesla believed he could top Marconi, both in the distance he could bridge and the energy he could transmit. In the spring of 1899, Tesla boarded a westbound train and headed into this wireless future.

...

WHERE BRAHMIN BOSTON PRIDED itself as a hub of learning, industrial Chicago preened as a center of commerce, and no edifice better symbolized the city's aspirations than the massive Auditorium Building. The great hulk of granite and steel looked across Michigan Avenue to the bustling railyard and lakefront beyond, and its boxlike exterior enclosed a veritable metropolis: offices and shops, a four-hundred-room hotel, a cavernous theater, and—up on the seventh floor—a banquet hall with gilded chandeliers. In such a space at such a time, paunchy gentlemen who chewed cigars and raised toasts gathered to feast sumptuously, and it was here that Chicago's capitalist elite—bankers, railroad tycoons, manufacturers, merchants—

THE AUDITORIUM BUILDING AND LAKE FRONT PARK.

assembled on a pleasant Saturday evening in May. They came to hear an invited guest, introduced to the room as "the magician of the electrical world."

Tesla stepped forward. With deep eyes, angled cheeks, and a mystical elegance, he did look the part of a wizard with a touch of the devil incarnate thrown in. ("Only the addition of a pointed beard would be needed to turn [him] into a typical Mephistopheles," remarked one who met him.) The air of mystery heightened when he began a demonstration to show what wireless could achieve. Within the ornate banquet hall, its air suffused with the aroma of beef tenderloin and roast squab, an artificial lake had been constructed at Tesla's request. On the shallow water floated what looked like a small submarine with an antenna sprouting from the top. Tesla claimed, astonishingly, that he could control the vessel without physical contact.

"The power is stored within the boat by means of a storage battery," he explained in refined English tinged by a Serbian accent. "We thus have the motive power. How to supply the mind? By transmitting the sensibility of my own mind to the boat. How shall we do this? Thus."

Tesla stood some twenty feet away, behind a white-clothed table upon which sat a small switchbox. He turned a lever, and the boat's propeller began to whir. The audience broke into applause. Tesla calmly smiled as he steered the vessel, wagging its rudder to starboard and port. The crowd gazed, spellbound. Although the robotic craft looked like a toy, it was in reality a scale model of a torpedo boat that Tesla was proposing to the U.S. military. His idea, inspired by America's war with Spain, was to create a crewless ship—a drone, controlled from a distance. To cap his demonstration, he transmitted another wireless signal to simulate the firing of the weapon. A shot rang out as smoke and flame issued from the bow.

Tesla elaborated on his demonstration the following day.

"The invention of the torpedo boat worked at a distance and the discovery of the principles of wireless telegraphy are simply incidents in my progress. They are not ends which I deliberately set out to accomplish," he told a reporter. "I want to go down to posterity as the founder of a new method of communication." Asked to amplify, Tesla offered an example. "Signaling to Mars?" he said. He leaned forward, his eyes fixed with intensity. "I have apparatus which can accomplish it beyond any question. If I should wish to send a signal to that planet I could be perfectly certain that the electrical effects would be thrown exactly where I desire to have them and that the exact signals I desire to make would be made. Further than that, I have an instrument by which I can receive with precision any signal that might be made to this world from Mars."

The yellow press seized on the news, giddy to revisit two of its favorite subjects: the red planet and the mysterious Nikola Tesla. MAYBE WE CAN TALK TO MARS, Hearst's *New York Journal* declared as it disclosed Tesla's news from Chicago, pairing that article with a second piece: GARRETT P. SERVISS TELLS HOW MARS MAY BE SIGNALLED. Here, the respected science journalist offered practical thoughts on how wireless telegraphy might be used to establish contact with aliens. Where Camille Flammarion had suggested an interplanetary code of communication by means of light signals arranged in some geometric shape—say, an equilateral triangle—Tesla was suggesting using *radio* signals arranged meaningfully in *time*. Serviss explained how Tesla might do this. "Let him send three dots or dashes in rapid succession and after an interval another succession of dots and dashes," Serviss said. "An intelligent person on Mars would see somebody was trying to talk with him. . . . If Mr. Tesla should persist in following up the telegraphing, in time the receiver on Mars would undertake to find out where it came from." The plan worked the other way around too. If the Martians wanted to call us up by radio, they might well adopt the same scheme.

With news of his Mars comments swirling in the papers, Tesla prepared to depart Chicago, but he was not heading home to New York. He would first continue west to the Rocky Mountains.

"Are you going to Colorado for a rest?" a reporter inquired.

"Rest?" the inventor replied. "Oh, no."

. . .

TESLA SOON ARRIVED IN the prim little city of Colorado Springs, its tidy homes connected by electric streetcars and a telephone system. Here on the plains south of Denver—where thundering buffalo had a few decades earlier sustained bands of the nomadic Ute, Arapaho, and Cheyenne tribes—settlers from the East Coast and from Europe had now taken up residence, many of them tuberculars drawn by the "light, dry, electrical atmosphere" that was said to be good for the lungs. (Such staticky conditions, however, made it "not a good country for hair," Tesla complained.) The well-groomed inventor checked in at the Alta Vista Hotel, four stories of pressed brick served by an elevator, then occupied a room with a view toward Pikes Peak. The great mass of pink granite loomed to the west, rising to a treeless, sculpted summit. Tesla's eccentricities quickly became evident to the hotel staff. The inventor exhibited symptoms of obsessive-compulsive disorder—a phobia of germs and an urge to repeat his actions by numbers divisible by three. From housekeeping he requested eighteen clean towels a day. "If he wants them, let him have them," the hotel manager's wife told a maid. "It is good to have a clean gentleman in the room."

To the east of town, on a high piece of prairie where cattle grazed, Tesla erected a peculiar structure. The building itself was crude and barnlike, with a large double door. An antenna crowned by a metal sphere emerged through the roof. This was Telsa's new laboratory, and, to keep prying eyes away, he padlocked it and posted warning signs—GREAT DANGER – KEEP OUT—though one observer did get inside and reported it "filled with dynamos, electric wires, switches, generators, motors, and almost every conceivable invention known to electricians." At night, the laboratory flashed and boomed as giant sparks flew. Townspeople observed the inventor's doings with apprehension. "We all thought he was crazy," recalled the motorman on a local streetcar line.

When reporters asked Tesla what he was up to, he evaded. "Of course you cannot expect me to expose my plans," he told one journalist, although he added with assurance: "No, I have no intention of communicating with Mars." For eight long months—in summer rain, autumn sun, winter snow—Tesla explored his ideas for transmitting electricity through the earth and atmosphere. He often used his wireless receiver to track electromagnetic signals from distant thunderstorms. With each stroke of lightning, the device would emit a *click* or sound a tone, as if the Earth itself were speaking. Alone in his laboratory one night, listening to what he termed "the pulse of the globe," he noted a feeble signal. *Click-click-click. Click-click-click.* It repeated in triplets, like a secret code. *One-two-three. One-two-three.* Though barely audible, the sounds drilled into his mind. He puzzled over their meaning. He told no one about them.

EXPERIMENTAL STATION AT COLORADO SPRINGS.

...

MEANWHILE, ACROSS THE BROADER nation, if not the world, the sense of foreboding that had accumulated toward the century's end was beginning to lift. William McKinley, on a visit to Chicago in the fall of 1899, spoke to the same businessmen in the same banquet hall that had hosted Tesla in the spring. "We have had a wonderful industrial development in the last two years. Our workshops never were so busy," the president boasted. (Some of those workshops—those factories—produced a new marvel, the automobile, a direct outgrowth of the bicycle.) "We have everything, gentlemen, to congratulate ourselves over as to the present condition of the country." To be sure, the troubles of the late-nineteenth century were not over—within two years, McKinley would be shot dead, yet one more head of state felled by the wave of anarchist violence—but the economy was, at last, emerging from its long depression. So, it seemed, was Percival Lowell.

More than two years had passed since Lowell first fell ill with neurasthenia, and he was now recuperating at another seaside resort favored by moneyed invalids, along the French Riviera. At the Grand Hôtel Costebelle in palm-fringed Hyères, during languid days of solitary walks and lunches with friends from Boston, he plotted his return to astronomy. Before this latest trip to Europe, Lowell had developed a warm professional relationship with another Massachusetts stargazer, David Todd, who ran the observatory at Amherst College. Todd had authored a new textbook that dedicated several pages to Lowell's Mars theory. "The explanation of the canals themselves . . . seems very plausible," Todd had written. "Of course, acceptance of this theory implies that Mars in ages past, has been, and may be still, peopled by intelligent beings." Lowell took an instant liking to this fellow astronomer who evinced a willingness to defy scientific orthodoxy. "Todd . . . is not a fossil but belongs to a living species," Lowell judged.

David Todd's wife was no fossil either. Mabel Loomis Todd was a spellbinding lecturer and prolific writer whose life intersected in intimate ways with another prominent family in Amherst, the Dickinsons. She had befriended the reclusive older sister, Emily, whose seemingly strange poetry she admired, and after Emily Dickinson died in 1886, it was Mabel Todd who first transcribed, assembled, and published the poems to world acclaim. Meanwhile, with the full knowledge

DAVID AND MABEL TODD,
AMHERST, MASSACHUSETTS.

and assent of her own husband, Mabel Todd engaged in a passionate affair with Emily Dickinson's brother, Austin, himself married. The thirteen-year relationship aroused constant gossip in Amherst and ended only with Austin's death.

Mabel Todd cast a mighty spell with her brown eyes and irrepressible energy. "I cannot explain it, but I feel strength and attractive power, and magnetism enough to fascinate a room full of people," she had written at age twenty-five, an early entry in a lifelong string of diaries and journals filled with astute observations about herself and others. "Mr. Percival Lowell . . . is very charming," she noted when the Boston astronomer visited her home in 1897. At that time, she had recently written a book on a favorite astronomical subject—*Total Eclipses of the Sun*—and she presented a copy to Lowell, then still in the throes of his depression. "Your book is on my table," he wrote in thanks. "I can only hope that the next time I have the pleasure of seeing you I shall not even be under a partial eclipse."

As for the Todds' unconventional marriage, it was loving, supportive, and full of adventure. "We spend our lives chasing eclipses," Mabel Todd told the Boston journalist Lilian Whiting, for every few years her husband organized another expedition to another exotic locale to catch another fleeting conjunction of the sun and moon. Total solar eclipses offered David Todd a chance to study Earth's home star, while his wife accompanied him to gather material for her lectures, articles, and books. The Todds cultivated wealthy benefactors to underwrite these costly expeditions, and their latest patron was

Lowell. While the neurasthenic astronomer relaxed on the Riviera in the winter and spring of 1900, his money was hard at work across the Mediterranean, in North Africa, where David Todd used it to prepare telescopes and cameras in advance of an eclipse that would occur at the end of May. Four days before the event, Lowell arrived on the scene to join the expedition. He seemed a changed man. "Mr. Lowell quite jolly," Mabel Todd reflected in her diary. Lowell, meanwhile, noted of her: "She is here as ubiquitous and chattery as ever."

The afternoon of May 28 found the three friends from Massachusetts at their predetermined observation post, atop the British consulate in Tripoli. The broad, flat roof surrounded a square courtyard and offered a view across the city, the low skyline punctuated by slender minarets topped by golden crescents. At the port, toward the east, masted ships bobbed on cobalt water. As the day progressed, the usual glare on the whitewashed buildings strangely softened to a silvery light when the moon began to cover the face of the sun. Soon the sky darkened—to purple, to slate, to steel. Swallows emerged in the cool desert air, darting through aberrant twilight. Arabs and Turks, the residents of the city, crowded the streets with hushed excitement. Suddenly, the sun vanished, replaced in the heavens by a glorious halo that blossomed behind the jet-black moon, a sight that breathed "the very spirit of interplanetary space," as Mabel Todd described it, "where time is not, where nothing is old, yet never young, in presence of which mere human emotion fades and faints and utterly dies away."

The total eclipse proved transcendent. Though lasting just fifty-one seconds, it was like an electric jolt that resuscitated Lowell and his reputation, for he now once again was deemed an astronomer of note. "The day was splendid," *The Boston Daily Globe* reported in a cable dispatch from Tripoli. "The eclipse expedition, under Prof Todd of Amherst college and Percival Lowell of Boston, completed successful observations."

Lowell had emerged from his own personal darkness, yet Mabel Todd could tell that not all was light. "Mr. Lowell is better, decidedly, than when he was in Amherst, and most of the time he is thoroughly charming—nobody can be more so," she wrote in her journal, "but occasionally there are irresponsible bursts that show a diseased mind."

...

SEVEN MONTHS LATER, AN epochal transition loomed for the entire planet as the nineteenth century approached its end. "This threshold of a new year and of a new century—such a moment between the past and the future—is one singularly rich in suggestion as well as in reflection," the ever-buoyant Lilian Whiting wrote as she assessed the monumental era just past. "That men would go about in carriages without horses, propelled by a force invisible to the eye; that they would send messages through the air . . . would have been regarded by those living at the opening of the nineteenth century as strange and supernatural possibilities." One could feel the anticipation of marvels as yet unknown, of a new era about to start.

The exact moment at which the twentieth century would begin, this razor line of history, had been fiercely debated for years. One side placed the transition at the end of 1899, the other at the close of 1900. This latter view won the day, with Flammarion leading the charge. "At midnight of December 31, 1900," he wrote, "will the hour glass of the nineteenth century run out and at the next moment the twentieth century will start on its career." New Year's Eve at the close of 1900 proved a night both memorable and exceptional. "Let the people of the other worlds know that on this planet a remarkable event is being celebrated," Hearst's San Francisco *Examiner* implored its readers, and much of America seemed to heed the message. Church bells echoed across Boston, steam whistles screeched in Chicago, and percussive fireworks filled the sky over all the major cities while boisterous crowds cheered and jostled in the streets below.

Meanwhile other, more subdued celebrations took place in churches, theaters, and civic halls. These solemn "watch meetings," held as fundraisers for the American Red Cross, featured patriotic music, speeches, and—the highlight—the recitation of letters from world luminaries who had been asked to reflect on the progress of the last century and the promise of the next. Among the celebrities who had sent their greetings were Queen Victoria, in her final weeks of life; Theodore Roosevelt, now vice president–elect; Susan B. Anthony, the suffragist; and that ever popular astronomer, Camille Flammarion. Most of the letters, when read aloud, elicited quiet reflection and polite applause, but one would explode upon the public at the very moment the new century was born:

> To the
> American Red Cross
> New York City
>
> The retrospect is glorious, the prospect is inspiring: Much might be said of both. But one idea dominates my mind. This—my best, my dearest—is for your noble cause.
> I have observed electrical actions, which have appeared inexplicable. Faint and uncertain though they were, they have given me a deep conviction and foreknowledge, that ere long all human beings on this globe, as one, will turn their eyes to the firmament above, with feelings of love and reverence, thrilled by the glad news: "Brethren! We have a message from another world, unknown and remote. It reads: one . . . two . . . three . . . . ["]
>
> —Nikola Tesla

Part Two

A NEW CIVILIZATION

1901–1907

CHAPTER
9

Messengers

1901

On the first morning of the new century, in a headline, Hearst's *New York Journal* asked the question that hung in the air: HAS NIKOLA TESLA SPOKEN WITH MARS? The enigmatic inventor soon answered.

Leaving the Fifth Avenue opulence of his home at the Waldorf-Astoria, Tesla headed thirty-three blocks south to the grittiness of his laboratory on East Houston Street. Tenements sat wedged between the industrial buildings, the narrow apartment blocks crammed with immigrants who lived and toiled in dark, squalid, hazardous conditions. Millions of these new Americans were arriving from Eastern and Southern Europe only to find themselves and their children earning pennies in sweatshops, at pushcarts, or within their own crowded and claustrophobic flats. Tesla—a proud immigrant himself—navigated the frigid streets, past horsecars and trolleys, to arrive at his workshop. It occupied two floors of a nondescript building, the space filled with the tools of his craft: vacuum tubes, oscillators, transformers, motors.

Beset by journalists who clamored for the meaning of his cryptic New Year's Eve message, the sorcerer emerged to address their queries. His face and hands were smeared with grease. Acid-eaten overalls draped from his lanky frame. "I would have abstained from making these observations known for some time yet, had I not been asked by the Red Cross Society to give a short expression of opinion for their meetings," Tesla said, feigning reluctance to explain.

Yet explain he did. "I went to Colorado early in May, 1899," he began, then told the story of his experimental laboratory and the "feeble electrical disturbances"—*one-two-three, one-two-three*—that he detected one night that summer. "What could they be?" he recalled wondering. "Their character showed unmistakably that they were not of solar origin," he judged. "Neither were they produced by any causes known to me on the globe. After months of deep thought on this subject I have arrived at the conviction, amounting to almost knowledge, that these movements must be of planetary origin." The electrical impulses, arriving with metronomic regularity, were just the sort of signal he had imagined aliens might use to shout hello across the solar system. Hence his conclusion: "Inhabitants of Mars, I believe, are trying to signal the Earth."

If public reaction to the triangle on Mars in 1892 had boomed like a fireworks display, what resounded now was a thunderous blast like the sinking of the USS *Maine*. The yellow press, both the papers in New York and their copycats across the country, trumpeted the news as they had the explosion that sparked America's war with Spain. A banner across Hearst's San Francisco *Examiner* cried: WORLD SPEAKS TO WORLD WITH MYSTERIOUS SIGNALS THROUGH VAST SPACE. The Philadelphia Inquirer declared atop its front: TESLA THINKS HE CAN TALK WITH PLANETS. A full-page Sunday feature in *The New York Herald*— boldly headlined WILL THE DAWN OF THE NEW CENTURY OPEN COMMUNICATION BETWEEN MARS AND THE EARTH?—recalled Mme. Guzman's legacy in her will: 100,000 francs "to be given to any person of any nationality who . . . should discover a means

of communicating with a star or planet and receive a reply." Was Tesla about to win the Pierre Guzman Prize?

Adding fuel to the conflagration was news that arrived from Flagstaff. The red planet was once again nearing Earth—Mars was approaching its biennial opposition—and A. E. Douglass, the young astronomer who had originally scouted the location of the Lowell Observatory and was now running the operation in his boss's neurasthenic absence, had been watching through the telescope. In December, he had spied a bright light shining above the planet's dark edge, and while he assumed it was merely a Martian cloud (and repeatedly said so), the papers disgorged their usual vomitus about such things. "Are these lights the result of further efforts by our Martian cousins to communicate with us?" the *Los Angeles Daily Herald* conjectured. "Garrett P. Serviss, the popular astronomer, thinks enough of the suggestion to discuss the probable means that Martians would take to make us understand their purpose." (Serviss's influence on the public mind had by now grown enormously. His columns, published and syndicated by Hearst, reached millions.) Serviss's colleague, the editor Arthur Brisbane, joined the speculation. "Of this you may be sure," Brisbane wrote in the *New York Journal*. "Sooner or later we shall communicate with all the planets."

The more sober newspapers downplayed the uproar. *The New York Times*, in a bid to distance itself from the yellow journals that published anything, no matter how lurid or ridiculous, had recently adopted the slogan "All the News That's Fit to Print." It evidently considered the latest Mars news unfit, for it printed very little. *The Wall Street Journal*, conservative and no-nonsense, similarly ignored the ruckus. Yet even the most sensationalistic papers did not insist that the Martians were *really* calling. Part of what made the story so engaging, and worth following for weeks, was the spirited controversy that Tesla's bold claims provoked.

Guglielmo Marconi, just then conducting wireless experiments in the South of England, dismissed his rival's brazen conclusions. "I should attribute the alleged signals from Mars to local disturbances in the atmosphere," he told a reporter. "In earlier experiments, before my apparatus was perfected, I often received signals apparently from nowhere." Edward Holden, the Lick Observatory's deposed director—now scraping by as a freelance writer yet haughty as ever—snickered. "Until Mr. Tesla has shown his apparatus to other experimenters and convinced them as well as himself, it may safely be taken for granted that his signals do not come from Mars." Several astronomers dismissed Tesla as a "visionary," a term that at the time held a decidedly negative meaning: one prone to illusory visions. Other attacks were even more vicious and personal. "It is intimated by one or two who know him that Tesla . . . has never been the same man since he passed through his own body the awful currents which ordinarily are supposed to result fatally," *The Chicago Tribune* reported. In other words, Tesla might have fried his own brain. "The suggestion [is] that the great investigator has been incapacitated from sound scientific work for the last two or three years by the mental effects of the terrible experiments to which he subjected himself."

Tesla did not stoop so low when coming to his own defense—he attempted to stay professional—but a few friends and admirers punched back on his behalf. Especially pugnacious was Julian Hawthorne. Famed as the son of Nathaniel Hawthorne and a writer like his father, this journalist of yellow reputation gleefully mocked Tesla's detractors:

> All the tame beasts with long ears in the stables of science begin waving those ears vigorously and braying forth indignant scoffs and denials. Yes, so has it ever been, and will be. How eagerly will every so-called son of science who has the power of absorbing, but not of assimilating or of creating, rush to trample under his hoofs the man of genius. . . . Tesla, whose brain, compared with those of most of his contemporary scientists, is as the dome of Saint Peter to pepper-pots, had been trained to the hour, and the signal was not in vain. When and how will he reply to it?

Before long, a fictional Tesla did attempt a reply, in a breathless serial called "To Mars with Tesla" that ran in a weekly boys' magazine. It fea-

tured all the gimmicks of adventure romance: kidnappings and escapes, quicksand and desperadoes, a secret cypher and a daring young protagonist who strove to help Tesla signal the red planet. This bold youth at one point in the story uttered words that surely resonated with most any reader in 1901. "Mars!" the boy cried. "Mars, again; Mars everywhere, and all of the time!"

...

IF YOU WERE TO spend weeks immersed in the periodicals of this era, as I have done—browsing digitized archives and scrolling through microfilm at the Library of Congress—you would be struck by the sense of giddy optimism, of limitless possibility. Experimental aircraft were making tentative hops off the ground, and an automobile soon rocketed an astonishing seventy miles per hour. The mere phrase *twentieth century* conjured unimaginable progress, an age "of promise and hope" that the papers welcomed with souvenir editions and color supplements forecasting the marvels to come: a time when people will "drift through space on flying machines" and will telephone "from San Francisco to London, or, in fact, to almost any other portion of the globe." *The Ladies' Home Journal*, in a feature headlined WHAT MAY HAPPEN IN THE NEXT HUNDRED YEARS, looked forward to strawberries as big as apples, rose blooms the size of cabbages. The idea of talking to Mars, therefore, seemed just one more incredible prospect that might actually come true. "The proverb that there is nothing new under the sun is incorrect," a San Francisco rabbi apprised his congregation. "Things absolutely new are being discovered constantly."

The United States was renewing itself, changing rapidly. Although the nation remained deeply fragmented—the South's Jim Crow laws mandated separation of the races, Native Americans had been forced onto reservations, a chasm divided the moneyed elite from the slum-dwelling poor—yet much of the country was growing more socially cohesive. Large numbers of Americans read the same periodicals, purchased the same consumer brands, and sought out the same forms of entertainment. Ragtime, based on syncopated African rhythms, had become the musical rage. Great masses of workers and a growing middle class crowded the amusement parks, burlesque theaters, and dance halls that offered a boisterous escape from the drudgery of life. A more unified popular culture spanned the nation, and it was this space that the Martians now invaded.

Much of what shaped consumer culture was advertising, dominated by professional agencies that crafted sophisticated ad campaigns. The eye-catching copy filled magazines, plastered the walls of buildings, emblazoned streetcars, and notably featured Mars itself as a pitchman. "Send us up some Pears' Soap" read the message beamed down from the planet in a prominent ad for a familiar brand. Another ad depicted Mars as a hobbled globe, propped up by a cane. "Not feeling well, getting old and weak," the planet shouted to Tesla through a megaphone while requesting fifty cases of a patent medicine called Vin Palmette, "the Greatest Tonic Restorative and Strengthener in the World." Yet another advertisement had the Martians calling for quite a different sort of tonic: "Please send one bottle Bailey's Pure Rye."

THE FIRST MESSAGE FROM MARS.

The craze infiltrated ballrooms and concert houses. Dance orchestras from Scranton to Buffalo to Duluth performed a new two-step called "A Signal from Mars," a "catchy air, which is taking at a rapid rate with every one who has heard it," a New Orleans paper reported. Meanwhile, on the vaudeville circuit—amid the comedians, minstrels, ventriloquists, and trained animals—a troupe of diminutive actors debuted a skit called *A Peep at Mars*. The female star, just thirty-two inches tall, was none other than

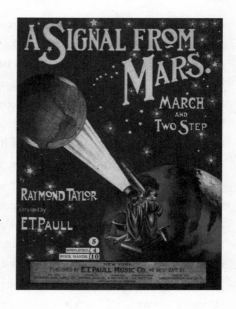

Lavinia Magri, better known as Mrs. Tom Thumb even long after the death of her first husband, whom she had married in an 1863 wedding spectacular choreographed by P. T. Barnum. Now performing opposite Primo Magri, her new husband and an actor of similar height, Lavinia played the role of an ingenue on Mars. Midway through the sketch, when bright lights flashed, she shrieked and clutched her brave suitor. "Be not afraid. I am with you," he consoled her. "The Earth is signaling Mars from Arizona. That's all."

Then, in October 1901, Horace Parker—a self-indulgent and self-satisfied Englishman—sat by the fire on a snowy night. He had thoughtlessly offended his fiancée, who left in a huff, but Parker seemed little bothered;

THE LITTLE LADY.

rather, he was content to enjoy a quiet evening with a glass of whiskey and a magazine. Just after 9:00 P.M., he opened the new issue of the *Astronomer* and perused its contents in the glow of a large standing oil lamp. "Well, now for Mars," he said as he began to read aloud. "'Latest observations have revealed strange lights which some astronomers believe to be signals put out in the hope of attracting an answer from our planet.' I don't believe a word of it," he muttered, yet the possibility of Martian life did intrigue him. Parker was a man of means and of learning, and he fancied himself a scientist. "If Mars is inhabited, I wonder what people are like," he mused. "Are they savages or are they ahead of us?" The light flickered as Parker drifted toward sleep, when a sudden clap of thunder roused him to consciousness. Opening his eyes, he found before him a man with tea-green skin dressed in the garb of a sheik. "Hullo, who are you?" Parker drowsily inquired. The stranger answered: "I am a Messenger from Mars."

Laughter and applause resonated within the confines of New York's Garrick Theatre, its elegant interior ornamented with plasterwork and stained glass, for this was the Broadway opening of *A Message from Mars*, a stage comedy that had transferred from London. The audience gleefully watched as the Martian emissary tried desperately to achieve his mission—to instill altruism in a most selfish human—while Parker, who chummily addressed his visitor as "Marsey," remained oblivious to his own obvious faults. The wildly successful play, which would enjoy long runs in New York and across the United States, exposed the cultural depths to which the Mars craze was penetrating. It also revealed, through its frivolousness, that the public in 1901 still

HE MEETS THE MAN FROM MARS.

viewed the existence of Martians as an imaginative flight of fancy, a topic more for entertainment than for sober contemplation.

In Act 2 of the play, Horace Parker, much like Ebenezer Scrooge in *A Christmas Carol*, begins to comprehend his own vanity with the help of his extraterrestrial visitor. To learn what others say of him in his absence, he is compelled to eavesdrop on a party across town. Parker's lawyer is there and suggests that his client might "turn his scientific abilities to some good purpose." "I'm afraid he can't do that," a prominent astronomer, Sir Edward Vivian, interjects. "My Dear Sir, his science is all fudge. Very praiseworthy in a wealthy man, of course. That sort of thing has to be encouraged among the wealthy, it is good for trade. But as far as any practical value—why the thing is simply absurd. . . . He has some fantastical idea about life on the planet Mars. Now all scientific men of any standing are quite agreed on this point. There *is* no such thing as life on the Planet Mars."

...

IN THE REAL WORLD, another wealthy amateur with fantastical ideas about life on Mars might well have wondered what others said behind *his* back. Not that Percival Lowell was unaware of his lowly standing in certain circles. He knew that some astronomers laughed at him after the Venus fiasco, and a hint now came that his Mars work was viewed derisively by some in his elite Boston milieu. In early 1901, a new book—written pseudonymously by the wife of an old friend—fleetingly lampooned both him and *The Atlantic Monthly*, which published his work on the Martian canals. Early in the novel, a Boston belle toys with the ego of a young Harvard graduate by pretending to mistake him for someone else. "Why, it was you, was it not," she says, "who wrote that splendid article in the last 'Pedantic' upon the occult influence of Mars?"

Now that Lowell had recovered from his nervous break, he was intent on resuming his astronomical work and resuscitating his reputation. As the new century's inaugural winter thawed into spring, he boarded a train to Arizona for the first time in four years. Stopping in Chicago at a hotel across from the Auditorium Building, where Tesla had made news on his own way west, Lowell distanced himself from the inventor who had sparked the latest Martian excitement. "I believe that Mars is inhabited," Lowell told a reporter, "but I do not believe the atmospheric disturbances which Nikola Tesla observed came from it."

Arriving back in Flagstaff, where the sheriff was trying to hunt down five outlaws who had escaped from the county jail, Lowell quickly positioned himself as the unrefined outpost's reigning intellectual. He hosted distinguished visitors: a paleontologist with the U.S. Geological Survey, a historian from Cornell, an old friend who taught English literature at Harvard. On July 4, when the town dressed itself in bunting and marched a parade down dusty streets, it was Lowell whom the citizens asked to give the oration at the foot of Mars Hill. When Lilian Whiting came through the area, the Boston journalist reported that "Mr. Lowell's work makes Flagstaff a scientific center of cosmopolitan importance." Lowell was back among admirers. "I am so much at home here," he wrote one of his sisters. He was unaware, however, of what was being said out of earshot.

Lowell's chief assistant, A. E. Douglass, had been young and subservient when first hired to help found the observatory. Now in his mid-thirties, he had grown independent and self-assured and had been running the facility himself for several years. "I need to talk to someone who understands scientific troubles," Douglass wrote his old Harvard colleague William H. Pickering, who had originally conceived the idea of the observatory in Arizona. Douglass explained that he was distressed about his boss's return to Flagstaff. "It appears to me that Mr. Lowell has a strong literary instinct and no scientific instinct," he complained. "I had supposed that he was anxious to acquire the latter and nothing could have given him a stronger incentive than this Venus business," Douglas continued, but he saw no evidence that Lowell had learned from his mistakes. "I wanted to say something of this kind to somebody," Douglass vented, "and there was no one else that I would say it to."

The following week, Douglass did find someone else to say it to: Lowell's brother-in-law (and cousin), William Lowell Putnam. "I am deeply attached to Mr. Lowell and would like to see his name have the highest and best renown," Douglass wrote, portraying himself not as a critic but as a well-intentioned messenger. He explained that if Lowell aimed to acquire widespread respect, he needed to address some shortcomings in his scientific method. Specifically: "carelessness . . . looseness of generalization verging toward misstatement . . . [and] neglect of alternative hypotheses." For example, Douglass wrote, "[Lowell] does not consider seriously any other hypothesis of the canals than the habitation hypothesis." What Douglass correctly diagnosed in his boss was a reckless combination of haste and narrowmindedness. "His work is not credited among astronomers because he devotes his energy to hunting up a few facts in support of some speculation instead of perseveringly hunting innumerable facts and then limiting himself to publishing the unavoidable conclusions, as all scientists of good standing do." Douglass urged Putnam to employ some gentle persuasion on his brother-in-law, to encourage less impetuousness in his astronomical work. Douglass concluded with a postscript: "Please consider this letter as between ourselves only."

Douglass must have known that the letter was a powder keg; he waited twenty-four hours before mailing it. Upon receiving it, Putnam did keep the contents confidential for several months, but, as one might have expected, he eventually divulged what Douglass had said. On a Friday in early August, Lowell—in Flagstaff—sent for his faithful assistant. Douglass stepped forward and was summarily fired. Lowell offered no explanation.

Blindsided, Douglass struggled to right his scientific career. It would take five years, but he finally reemerged as a working astronomer when he joined the faculty at the University of Arizona. As a professor, he would go on to found a new observatory and to pioneer the field of dendrochronology—the study of tree rings. It was my good fortune that Lowell's discarded assistant made such a name for himself, for Douglass's papers today remain safely stored at his old college campus in Tucson. There, in a box of correspondence, I found the carbon of a letter Douglass composed in 1937—long after Lowell's death—in which he described the circumstances of his firing and its impact. "The events occurred many years ago," he wrote with the perspective of a then seventy-year-old. "It was a turning point in my career for it made me depend very much on myself." The letter, it seems, served as a sort of catharsis, for Douglass had written it to Lowell's younger brother, himself eighty. It was one man, toward the end of his existence, sharing with another a tale of the vagaries of life and career.

...

A SCIENTIFIC REPUTATION IS a fungible thing. It can rise or fall based not only on one's professional employment and deeds but also on one's affiliations, and Lowell—after dismissing Douglass from his observatory—soon benefited from his association with a most influential scientist from across the ocean. Sir Robert Ball taught astronomy at Cambridge University, wrote popular books, and was a perennial star of the lecture circuit. He was a jovial Irishman with a generous middle and one eye made of glass, and in the fall of 1901 he arrived in America for an ambitious speaking tour. Ball would give nearly fifty talks in eleven weeks. His first stop was Boston and the Lowell Institute.

Percival Lowell's father, who ran the institute for two decades, had recently died—leaving charge of its affairs to Percival's brother. (Their father's death also left both sons with a vast amount of additional wealth, which freed Percival to spend yet more time and money pursuing his Mars obsession.) When Sir Robert Ball came to town as their guest, the Lowell siblings entertained him with a dinner at the Back Bay home of Percival's sister, Katharine Roosevelt. They invited local sci-

SIR ROBERT S. BALL.

entists to attend, including Edward Morse, the naturalist whose lectures had inspired Percival's initial turn toward to Japan. "A pleasant night it was," Ball wrote. "There is certainly a most brilliant and cultured society of men and women in Boston." Ball counted Lowell's *Mars* among his favorite books, and as he barnstormed the United States, he repeatedly and publicly spoke of Lowell's theories.

Lecturing in Philadelphia, Ball displayed drawings of Mars and "paid tribute to the recent observations of Percival Lowell," a reporter noted. In New York, Ball discussed Mars with Garrett P. Serviss. "I do believe in the existence of the so-called 'canals,'" Ball said in the interview. "We have the observations of your Mr. Lowell and others, and there can be no doubt that the lines are there." Ball put his opinions in writing for Hearst's newspapers, which typeset the words as if the Irishman had shouted them: "The canals which astronomers have long observed on the planet Mars and which are believed to be rivers or canals ARE STRAIGHT LINES. It is an axiomatic truth in science that NATURE DOES NOT WORK ALONG STRAIGHT LINES. THESE CANALS, THEN, ARE THE WORK OF THE MARTIANS—the INHABITANTS OF MARS." Ball's message proved persuasive. An astronomy professor in Indianapolis soon announced that he had become "a convert to the theories advanced by Sir Robert Ball, Garrett Serviss and others, that the planet Mars is inhabited." A Massachusetts poet set the cosmic mystery to rhyme:

> Since man began the skies to scan
>   And at their secrets try to get,
> Of all the stars, the planet Mars
>   Has been the biggest riddle yet.
> The lines we trace upon its face,
>   The scientific sharps agree
> They can't explain, though straight and plain
>   And set by rule, as all may see.

Lowell and his small army of canalists were rallying, much to the chagrin of the enemy camp. As long as maps of the planet continued to depict a network of straight lines—so straight they appeared drawn with intention—it was clear that the Martians would live, at least in the human imagination. For the anti-canalists, the task at hand now was to erase those lines entirely, a campaign that would begin in the coming year.

...

MEANWHILE, ON A THURSDAY in mid-December, before 1901 drew to a close, Marconi—that man of wireless fame—braced himself on a blustery cliff above the coast of Newfoundland, a place known prophetically as Signal Hill. Into the teeth of the wind, he and his men released a giant hexagonal kite, which jolted skyward and carried with it a thread-like antenna. Once the swaying line reached a suitable altitude and had been secured to a pole, the inventor entered an adjacent building, where he had set up his receiving apparatus. Inside it was calm and quiet. Marconi pressed a telephone receiver to his ear. He strained to listen as the minutes passed. Finally, at 12:30 P.M., his equipment started to respond. *Click-click-click.* The signal was faint, but Marconi found it unmistakable. *Click-click-click.* It paused, then repeated at intervals—*one-two-three, one-two-three*—an invisible message from across the void.

SIGNAL HILL, ST. JOHN'S, FROM THE SEA.

CHAPTER
10

"Small Boy Theory"

1902–1903

In 1867, a full decade before Giovanni Schiaparelli first mapped canals on Mars, a popular book appeared in Paris. Called simply *L'Optique*, it described the human visual system—how it works, and how it fails. "Our very eye, the most remarkable and most important of our senses, which we deem so sure and so impeccable, is constantly deceiving us," the book explained. Indeed, the eye and brain often misinterpret what we see, and such defects fascinated scientists and artists in the late nineteenth century. Physiologists, painters, and architects found ways to fool the viewer into misperceiving colors, angles, brightness, and depth by manipulating visual cues. Such optical illusions could make straight lines appear bent, identical shapes seem of different sizes. "[There are] numerous means that art has invented to seduce the sense of sight, and to give it purely imaginary impressions," the book concluded. The author, per the title page, was identified as Fulgence Marion, but he, too, was an illusion. The true author was Camille Flammarion, who in his younger days hid behind a pseudonym when writing on non-astronomical matters to avoid diluting his brand.

By the time the new century began, inventors found new ways to put illusions to work. Newspapers published photographs using halftone printing, which translated subtle shades into an array of black dots that deceived the brain into seeing recognizable shapes: the Eiffel

Tower, a baseball player, Teddy Roosevelt's spirited grin. Pioneers of film such as Thomas Edison and the Lumière brothers projected still images in rapid succession to fool the eye into perceiving continuous motion, a trick that seemed pure magic. And while audiences gaped at moving pictures, they also marveled at stationary pictures that seemingly transformed themselves before the eye. In 1902, *Life* magazine, a periodical famous for its illustrations (and, in a later incarnation, its photography), published a clever and popular illusion that, when seen up close, depicted a young woman at her dressing mirror. From a distance, however, the image took a grim turn. Its details—the woman's head and its reflection, the bottles of perfume and cosmetics along the table—blended into the shape of a grinning skull, a reminder of the impermanence of life and beauty. The picture forced the brain to toggle between two interpretations and served as a visual and linguistic pun. Its caption: "All is vanity."

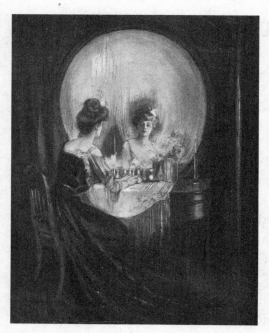
ALL IS VANITY.

A British astronomer suspected that Mars might be playing a similar trick through the telescope. "We cannot assume that what we are able to discern [on the planet] is really the ultimate structure of the body we are examining," argued Edward Walter Maunder, a longtime assistant at Britain's Royal Observatory in Greenwich. Maunder's tenure at that esteemed institution had begun in the 1870s, inauspiciously—his first boss assessed him "the veriest dummy that I ever saw"—but by the 1890s, Maunder had gained a solid reputation as an expert on sunspots. Those blemishes that occasionally freckle the solar disk are often too small to be seen individually without a telescope, but Maunder found that groups of sunspots could—like the dots of a halftone photograph—give the illusion of shapes, including lines.

Maunder's solar observations made him wonder about Mars. The Martian surface was unquestionably covered with an abundance of features—varied contours and colors, areas of stippling and shading. Maunder suspected that the eye and brain, trying to make sense of detail too small to be perceived, might impose straight lines on the chaos. Perhaps the Martian canals were nothing more than areas of speckling that the visual system morphed into linear form. Or maybe they were soft boundaries between zones of different color that the eye perceived as hard edges. "'Canals' in the sense of being artificial productions, the markings on Mars, which bear that name, are certainly not," Maunder wrote in 1895. Over the years, even as Percival Lowell's confidence in the canals grew, Maunder's doubts did not ebb. In 1902, now that the frenzy of speculation about intelligent Martians had reached unprecedented heights, Maunder chose to revisit whether those lines on the planet might result from a "canaliform illusion." He devised an experiment.

To conduct his research, Maunder had first to identify a large group of suitable test subjects; he needed individuals who knew little about Mars and could be counted on to follow instructions, no matter how odd. In this quest, the British astronomer found himself well situated. His workplace, the Royal Observatory, sat on a leafy hill just east of London and looked down across parkland to a sharp bend in the Thames. There, perched on the riverbank, sprawled the Baroque buildings of the Royal Naval College, which formerly served as a home for retired sailors called the Royal Hospital, and across the way stood the Royal Hospital School, which prepared the sons of seamen for careers in the navy. One could see the boys, age eleven to fifteen, as they drilled and marched across the grounds. These youngsters in their uniforms looked like just the sort of dutiful, intellectually naïve subjects Maunder needed. He approached the school's headmaster,

MARCHING TO THE DINING HALL.

and together they conscripted the students for a curious set of experiments that would occur over the following year.

...

MEANWHILE, A REINVIGORATED PERCIVAL LOWELL was putting his energies to constructive use. He restaffed his observatory, hiring two assistants to replace the one he had fired. The new men were talented but callow, therefore unlikely to challenge his knowledge and authority. Lowell ordered the construction of a comfortable home, tucked in the pine forest beside his telescope on Mars Hill. Though managed by servants and adorned with a music box in the dining room, the Baronial Mansion, as Lowell called it, was decidedly rustic. Walls of unpeeled logs offered refuge for insects. Pack rats crawled in the attic.

Lowell adroitly leveraged his wealth in other ways, to bolster his reputation. He had long served on the governing board of MIT—as did his brother, several cousins, and his late father—so when his father died, Lowell (with his siblings) gave the university a portion of the vast inheritance. The university in turn gave him a faculty title: nonresident professor of astronomy. The position was largely honorary and carried few responsibilities, but while it provided little or no compensation, it offered something more precious than a mere salary. It bestowed academic prestige.

The newly minted professor took the long train ride from Boston back to Flagstaff in early 1903. Just then, the red planet was once again nearing Earth, as Garrett P. Serviss excitedly told the public. "Anybody can see Mars now, climbing up the eastern sky before midnight, lurid among the white stars and slowly growing brighter as he approaches his place of opposition to the sun," Serviss wrote. "This time Mars shows us his northern hemisphere, and that is a circumstance calculated to pique curiosity, for when he furnished his great sensations in 1892 and 1894, it was his southern hemisphere that he turned in our direction."

From winter through summer in Flagstaff, Lowell studied the lesser-known northern half of the planet. That hemisphere was enjoying its own late spring and summer, and Lowell, watching through the telescope, sketched the changing scene. He saw the canals and oases appear, then darken, first in the north—near the polar ice cap. This was precisely as things should occur if his theory was correct that vegetation, fed by meltwater in the warmer months, was greening the red soil. "Mars is behaving in satisfactory accord with my understanding of it from previous oppositions," Lowell wrote his brother-in-law. Then, like God surveying His own creation, Lowell added playfully in French, "*Ce que je trouve bien.*" Lowell found it good.

...

AS FOR THE WIRELESS message received by Marconi in Newfoundland on December 12, 1901, it had not come from Mars, but that fact hardly made the news less worthy. "Guglielmo Marconi announced to-night the most wonderful scientific development of recent times," the papers gushed. "He stated that he had received electric signals across the Atlantic Ocean from his station in Cornwall, England." In this case the *click-click-click* represented *dot-dot-dot*—the letter *s* in Morse code—which Marconi had instructed his own men to flash repeatedly as a test of whether his technology could wirelessly connect the continents. It could and did, representing a quantum leap in the distance traversed by radio waves.

"[Marconi] has gained so enormous a step at a single effort that it seems hard to set a limit beyond which these strange impulses cannot reach," wrote Garrett P. Serviss, who soon sat down with the Italian inventor. "There is no reason why we should not send electric waves to the moon or to Mars, or any other planet, is there?" Serviss asked.

Marconi smiled as he answered: "Apparently, there is not."

The news did not make Nikola Tesla smile—Marconi had bested him—but Tesla aimed to leapfrog his competitor with a far more powerful wireless system, one that could reach not just over the ocean but through the bulk of the earth. To implement this scheme, he moved his operations from Manhattan to the north shore of Long Island, to a lonely beachfront being transformed in a summer resort community. Developer James Warden, who called the place Wardenclyffe, sold Tesla two hundred acres in which the inventor sowed his grandest dreams.

NOW IT'S UP TO TESLA TO CONNECT US WITH MARS.

What soon emerged from the soil was an enormous tower that loomed above the trees like a Martian tripod in *The War of the Worlds*. The latticework structure of stilts and struts, crowned by a domed cage, straddled a deep shaft in the earth. Tesla refused to divulge the purpose. "There is not one word to say about it, and I very much hope you will not mention it in the newspaper," he told a reporter, his secrecy serving only to breed speculation.

One seasonably mild evening in July of 1903, Tesla's strange colossus began to stir. "All sorts of lightning were flashed from the tall tower and poles last night," the New York *Sun* reported. "For a time the air was filled with blinding streaks of electricity which seemed to shoot off into the darkness on some mysterious errand." Wardenclyffe's vacationers, relaxing after a day of sailing and golf, enjoyed the show as if gazing at fireworks. "The villagers sit out in front of their houses, and at intervals between batting mosquitoes from their visages speculate on the meaning of the strange lights," continued *The World*. "Whatever Nicola Tesla is trying to do . . . he has succeeded in keeping the [locals] guessing. Some think he is trying to signal Mars."

A week later, however, the same paper raised doubts about whether Tesla could really have been conversing with the Martian canal builders. "At a recent meeting of the Astronomical Society in London," *The World* reported, "striking confirmation was obtained of the view that the 'canals' in Mars are due for the most part to an optical delusion." Walter Maunder, the expert on sunspots, had launched his attack.

. . .

THE GIST OF MAUNDER'S experiment, conducted at the Royal Hospital School, was to explore if he could trick the human eye into seeing nonexistent canals on maps of Mars. For each test, he began with an actual chart of the planet, in most cases one created by Schiaparelli or Lowell. Maunder left intact the large dark areas that most astronomers agreed were actual features of the Martian surface, but he erased the narrow canals and replaced them with more naturalistic shapes—winding rivers, for instance, or gentle speckling. The astronomer then brought this altered map to the school and hung it at the front of a classroom. He did not tell the students what the diagram represented—nor did they ask—but he

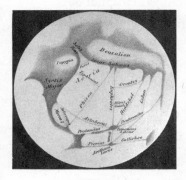

KEY MAP.

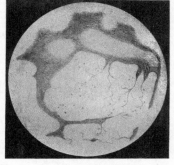

DRAWING SUBMITTED TO THE BOYS.

instructed them simply to copy it as faithfully as they could, "putting in all that they could see and nothing of which they were not sure." He forbade the boys to leave their desks; they could not approach the drawing for a better view.

The results, amassed in thirteen separate tests, were just as Maunder had predicted: many students saw, and drew, straight lines where none existed. At the end of one class, the boys crowded the front of the room. "There aren't any lines," one cried when he saw up close the diagram he had copied at a distance. Others erupted in confused amazement. "There are dots but not lines." "But I saw the lines." "I drew them all over, and there aren't any." The students imagined themselves the victims of a hoax.

"Just the same 'hoax' as [we] had worked off upon the Royal Hospital boys, the planet Mars has been working off upon his portrayers for the last quarter of a century," Maunder now publicly asserted both in print and in lectures. Maunder argued that the boys' brains perceived lines where their eyes saw marks too small to be distinguished individually, and this conclusion was bolstered by the fact that the students who drew the illusory "canals" were

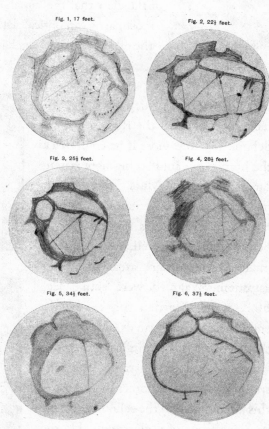

DRAWINGS MADE BY BOYS PLACED AT THE INDICATED DISTANCES FROM THE ORIGINAL DRAWING.

primarily those in the middle of the room. Those in the very front were close enough to discern the squiggles and dots that Maunder had added to the Mars map, and they drew those features with some accuracy. The boys in the far back, on the other hand, missed those faint details entirely and simply left large areas of their drawings blank. "[This result] explains why the 'canal' system as such was not recognised until a quarter of a century ago," Maunder concluded. "Earlier in the century the telescopes devoted to the study of Mars were mostly small, and the observers with them might be compared to the boys on the back row." More modern telescopes had, figuratively speaking, moved observers to the middle of the room, where the illusion became evident. One might hope that future observations with even larger instruments would be equivalent to advancing to the front row, where Mars's true surface would finally reveal itself.

When Maunder presented his findings before the British Astronomical Association, the assemblage chuckled. "[What] a great pity" to lose the canals, one in the audience chimed in sarcastically; "they served so many useful purposes—watercourses, lines of vegetation . . . motor-car tracks." Another reminded his colleagues how Percival Lowell had once reported "a most superb display of canals on Venus" that Lowell later acknowledged existed only in his eye. The crowd guffawed. The speaker was curious to know if Lowell now "had 'owned up,' to use an American expression, in the case of Mars."

One in the group did rise in Lowell's defense. "These experiments prove nothing whatever," the man argued, both at the meeting and in a letter to a popular science weekly. "Almost any figure could be so drawn that it would appear correct when seen from a distance, but would be found of quite a different character when a near view was obtained." He mentioned an optical illusion he had seen, similar to the one labeled "All is vanity," which "depicted two or three children playing with wineglasses; but when viewed from a distance of three or four yards, it appeared to be an exact representation of a full-sized human skull. It is obvious, however," Lowell's defender continued, "that it could not be argued from this case that every picture of a human skull must really be a picture of children playing with wineglasses, and it would seem equally obvious that sketches of Mars may be quite correct when they show continuous lines." In other words, while Maunder's experiment showed that illusions *might* explain the Martian canals, it could not prove that they *did*.

This objection notwithstanding, the discussion at the meeting ended on a decidedly anti-canalist note. The astronomical association's president congratulated Maunder on his research, which "really cut away the ground from under the feet of those who thought they had been able to prove that there were canals. The onus of proof," he said, "now [lies] upon those who thought the canals were there."

...

THE CANALISTS DID NOT see the situation as quite so dire.

Garrett P. Serviss, ever helpful not only in informing Americans of the latest Mars news but in telling the public what to make of it, downplayed the significance. "It does not seem . . . that the results of these experiments should be taken as conclusive, especially in view of the great definiteness with which Schiaparelli and Lowell and a few others have drawn the lines seen by them," he wrote reassuringly. "The fact also that conspicuous changes of a definite character have been seen to take place in many of the canals"—that is, their seasonal darkening—"militates against the assumption that they are illusory."

Lowell hit back in different fashion; he ridiculed Maunder's choice of research subjects. "Under certain circumstances it is quite possible to perceive illusory lines," Lowell acknowledged, "[but] the whole art of the observer consists in learning to distinguish the true from the false." In other words, his years at the telescope should count for more than the inexperience of Greenwich schoolchildren. The aristocratic Lowell then employed a suitably elitist analogy. "Because a small boy would certainly not distinguish Oolong and Soochong teas by their taste," he argued, "it [is] unsafe to assert that a professional tea-taster cannot." Lowell fully rejected what he derided as Maunder's "Small Boy Theory." As he later underscored in a note to Camille Flammarion, "It is not me who makes lines on Mars, it is the Martians."

As for Flammarion, he, too, was unbothered by the news from

London. "The theory expressed by Mr. Maunder, that the canals are myths and the labors of the past quarter of a century are a mere web of vain mysteries, may prove attractive for some who do not like the trouble of thinking," the Frenchman declared to an American journalist. "But I doubt that astronomy will be disturbed by it.... That we see something [on Mars] is a fact which can scarcely be questioned, and which can certainly not be disproved."

It was an oddly definitive statement for Flammarion. After all, his pseudonymous alter ego, Fulgence Marion, had cautioned readers in the book *L'Optique* not to trust their eyes, but the astronomer displayed no such humility about himself. This is a common failing among the intelligent beings of Earth. We are prone to see biases in others yet are blind to those same weaknesses in ourselves. In this way, illusions result not just from the optics of the eye and the circuitry of the brain but from the workings of the mind, a mysterious world whose contours—like those of Mars—were just coming into focus.

CHAPTER

11

THE MARTIANS OF EARTH

1903

NEW YORK CITY, ALREADY A GLOBAL METROPOLIS, GREW EVER busier, taller, and more interconnected. The colossal Williamsburg Bridge stood nearly complete, its great span linking the Lower East Side with Brooklyn, while workers burrowed a subway beneath the grid of Manhattan streets. Overhead, a new wedge-shaped marvel of a skyscraper, the Flatiron Building, offered a God's-eye view of the impressive sprawl. In a quieter area a few blocks away, near Gramercy Park—that tranquil, private enclave for the upper class—evening services were about to begin at a church frequented by the city's elite.

VIEW FROM THE UPPER FLOORS OF
THE FLATIRON BUILDING.

It was a mild Sunday in the spring of 1903, and a well-groomed young man entered through the main door. Little noticed, he quietly strode up the aisle, approached the altar, then solemnly bowed and turned toward the pews. "Listen, ye mortals of Earth," he proclaimed in a loud and cultured voice. "I am a messenger from Mars, the first ever sent by our glorious ruler to the mean inhabitants of this lowly planet." Stunned parishioners looked on as the unknown caller lifted a prayer book. "Hear the message," he boomed. "I will read it to you." As the self-professed Martian was about to continue, the sexton rushed forward and asked the visitor if he might finish the sermon outside. The man complied. He bowed again, walked slowly out of the church, and vanished into the night.

This incident was hardly unique. Later that year, a woman in St. Louis brought tidings from the red planet. Her Earth name was Sara Weiss, though she claimed to go by Gentola on Mars, a world known by its inhabitants as Ento. She said she had traveled there many times—not physically, but as a disembodied soul—and she wrote a book describing the wonders she had found: iridescent cities, flying machines, exotic flowers, and a people as righteous as they were handsome. Gentola's guide on these journeys, himself a spirit, asked her to relay a message to two scientific men of Earth—Camille Flammarion and Giovanni Schiaparelli—who were "inclined to believe that the Entoans (Marsians) have resorted to irrigation." "Gentlemen," the message read, "I assure you that the surmise is entirely correct."

Other earthlings were asserting their own personal connection to Mars. A man from Boston said his soul spent a month on the planet while his body was meditating at a Buddhist monastery in India. A British clerk insisted that he received telepathic messages from a Martian woman named Silver Pearl, who spoke "with a slight foreign accent." These claims did not evoke universal ridicule. Mystical experiences were common at this time, when spiritualism was enjoying its own popular craze. Séances drew enthusiasts with the zeal of a religious revival. Ouija boards and ghost stories enthralled the masses.

Belief in the paranormal, already widespread since the mid-nineteenth century, had gained credence from recent advances in physics. The discovery of X-rays and radio waves revealed a hidden world filled with unseen forces, which made plausible that some form of invisible energy might explain thought transference, clairvoyance, and similar phenomena. (As a prominent medium explained, "Telepathy is nature's wireless telegraphy.") To test the idea, scientists held

EXPERIMENTAL SÉANCE AT THE APARTMENT OF
CAMILLE FLAMMARION. RUE CASSINI, PARIS, 1898.

experimental séances. Dabbling in this effort were astronomers, including Flammarion and Schiaparelli, but a more concerted campaign was waged by psychologists.

Théodore Flournoy, a psychology professor at the University of Geneva, spent several years studying a Swiss woman who reportedly traveled to Mars when in a trance. She worked as a salesperson in a silk shop by day, claimed psychic abilities, and did volunteer duty at séances in her off hours. To protect her identity, Flournoy gave her an alias—Hélène Smith. Confident, with dark and intelligent eyes, she exhibited grotesque behaviors when she fell into a hypnotic state; she groaned, cried, laughed, genuflected, shivered violently, and lapsed into Martian tongue. *"Ané éni ké éréduté cé ilassuné té imâ ni bétiné chée durée,"* she uttered at a séance in the winter of 1897. She later translated to her native French: *"C'est ici que, solitaire, je m'approche du ciel et regarde ta terre."* ("It is here that, alone, I bring myself near to heaven and look upon the Earth.")

Professor Flournoy took notes at the sessions and analyzed the Martian language—its phonetics, grammar, syntax. He found it all too familiar. "The order of the words," he discovered, "is absolutely the same in Martian as in French." What is more, the alphabets of both languages comprised nearly identical sounds, and many Martian words, when compared with their French equivalents, contained the same number of syllables. (*Métiche* was Martian for *monsieur*; *médache* for *madame*.) Flournoy argued that Martian, "in spite of its strange appearance and the fifty millions of leagues which separate us from the red planet," was nothing more than "disguised French."

It was clear to Flournoy that Hélène Smith had never traveled to Mars, but he did not consider her a fraud; she made no money from her psychic performances, her trances appeared genuine, and she seemed convinced of the reality of her astral journeys. Those journeys, Flournoy concluded, were hallucinations concocted by her own mind to escape a dissatisfying life. This dreamlike adventure—what he called a "romance of the subliminal imagination"—was a phenomenon worthy of study in its own right. The unconscious mind was just then being explored by other scientists and medical doctors, notably Sigmund Freud, who plumbed the abyssal depths of the self, beyond the sunlight of our awareness, where he found repressed memories. Flournoy's Martian-speaking psychic showed that those dark places also held a vast reservoir of creativity that can weave convincing fantasies to trick us into believing falsehoods we crave to be true. A historian of psychiatry would later remark that this mythmaking aspect of the unconscious, the so-called mythopoeic unconscious, is a force that has wielded "terrible power—a power that fathered epidemics of demonism, collective psychoses among witches, revelations of spiritualists." It is, in other words, a power than can unleash delusions that spread from person to person: social contagions.

...

PERCIVAL LOWELL, FOR ALL his aristocratic savoir-faire, exhibited an astonishing lack of self-awareness. In the late spring of 1903, the people of Flagstaff once again called on his oratorical skills, this time to give a Memorial Day address. After the town's Civil War veterans marched to the courthouse, the upright Bostonian took the stage to deliver a speech that was part eulogy for the dead, part rallying cry for the living. Lowell asserted that in order for the United States to thrive, united, its peoples of different beliefs and backgrounds must learn to cooperate, and in making his point Lowell compared the body politic to an individual mind. "In any active brain ideas are constantly originating. They crowd and jostle one another for recognition," he explained. "Now it is a curious and instructive fact of psychology, not so old of recognition as already to have become trite, that any and each of these ideas is a force which, unless inhibited by another idea, will instantly and inevitably produce its own particular effect.... We see this with monomaniacs where one idea acquires such momentum as to bear down all others before it. For by constant repetition any idea gets stronger and stronger." Lowell suggested that such a singular focus was unhealthy—whether in politics, when one party or faction gains too much control, or in the mind, when one idea chokes off consideration of others. Yet this, precisely, described the workings of Lowell's brain.

As far back as a decade earlier, when Lowell still lived in Japan, his friends diagnosed in him a growing intellectual arrogance. "He has become more certain of everything," an acquaintance ranted. "I cannot stand . . . the way he has of jumping at some general idea or theory, enunciated as a theorem," after which he "selects and bends facts to underprop that generalisation." This was much the same complaint that Lowell's assistant, A. E. Douglass, later made—and that got him fired. As if to prove his own stubbornness, Lowell had once advised Douglass, "It is better never to admit that you have made a mistake."

Lowell now turned this arrogance and monomania toward the latest planetary observations.

During his months at the telescope in the first half of 1903, Lowell had made 372 separate drawings of Mars, depicting its surface features as they rotated in and out of view. Where he had previously focused on the geometric layout of the canals—how they connected to one another, and to the oases—he now concentrated on their behavior, how they intermittently materialized and faded through the seasons. The canals of Mars were more than mere lines; they were places, each with its own personality, its own name. Schiaparelli had labeled them in the same manner as he had other Martian features, sprinkling his network of waterways with references to history and ancient legend. Some canals he christened after rivers, both mythological and real—the Styx, the Euphrates, the Ganges. Others he gave the names of classical gods, monsters, and heroes—Ulysses, Titan, Gorgon, Cyclops. Lowell, searching for evidence to support his predetermined theory, cast his eyes across this evocative landscape and spun a new story that, much like Hélène Smith's imaginary travels to Mars, served as his own romance of the subliminal imagination. This tale unfolded in distinct parts: a sensational three-act drama of the mind.

The setting for Act 1 was a pair of adjacent canals, near the region of Utopia, that conjured visions of ancient Egypt. The Amenthes, named by Schiaparelli for the Egyptian underworld, ran due north–south. Beside it, springing from the same oasis but veering off at an angle, was the Thoth, named for an Egyptian god. When Lowell looked at his drawings of these canals from 1903, he discovered an odd phenomenon: During periods when the Amenthes became conspicuous, the Thoth faded to invisibility. Later, when the Thoth grew visible, the Amenthes disappeared. Lowell inferred from the seesaw behavior a profound explanation: The Martian farmers, struggling

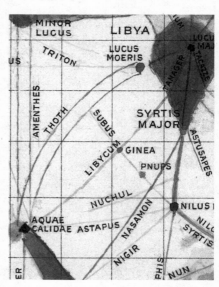

DETAIL FROM LOWELL OBSERVATORY MARS MAP.

to survive on a dry planet, were shifting the flow of precious moisture back and forth between agricultural districts. "It is easily conceivable that a limited water supply should involve a necessity of the sort," Lowell asserted in a scientific bulletin published by his observatory. "It may well be that after one district has enjoyed the water and its results for a certain period, the supply should then be turned for a time into a neighboring one to be turned back again after a while."

Act 2 in the drama proved yet more inventive. Lowell looked carefully at all of his sketches from 1903. He collated tables of numbers and plotted them on graph paper to show how each canal darkened or lightened over time. The result was more than one hundred separate curves of visibility. Lowell termed these *cartouches* after the ancient symbols representing the names of Egyptian kings, which when found on the Rosetta Stone permitted the long-sought goal of translating hieroglyphics. He hoped his own self-styled cartouches might be the key to solving the riddle of Mars. At first, the graphs were hard to decipher, a seeming jumble. "Each [canal] waxed and waned in a way peculiar to itself. . . . No connection appeared between their several behaviors pointing to a common cause." But when he grouped the canals by latitude and finessed the data, a pattern emerged—a wave of darkening that rolled down the Martian disk as the North Pole's ice cap melted in the late spring and summer.

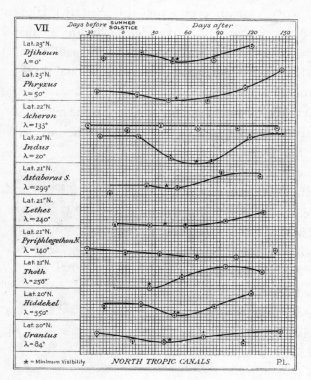

Lowell saw this as strong evidence to support his long-held theory that the canals transported polar meltwater and became visible as vegetation grew on the channels' banks, and his new analysis added stunning details. It revealed the rate at which water flowed: 51 miles per day, or 2.1 miles per hour. ("The mental ear detects the sound of water percolating down the latitudes," he later effused.) When he told his staff of the discovery, he was almost giddy. "I think I have found the law governing the development of the canal system."

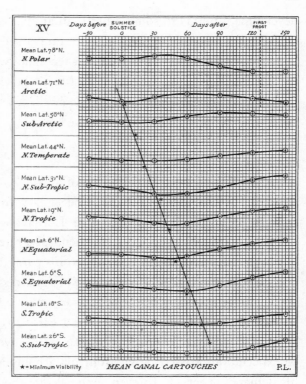

And there was yet more—the climactic third act in his drama. The wave of darkening, he wrote an assistant, moved "from the north pole to the equator and *beyond* south. The importance of this you will grasp: it shows that the development of [the canals] is dependent upon the melting of the polar cap and then proceeds down the disk not even stopping at the equator." Moreover, the same process seemed to occur in reverse a half year later. During the *southern* summer, when the South Polar cap melted, its water appeared to travel *north* past the equator, well into the Northern Hemisphere. This meant that the water in the canals flowed both ways.

Had Lowell been less intransigent, or less intelligent, he might have acknowledged a fatal flaw in his theory: *An irrigation ditch cannot transport water downhill in both directions.* Lowell, however, turned the flaw to his advantage. "Since there is no force to make water, say, flow from pole to equator [and back]," he wrote excitedly, "this law of development points to some other than any natural cause known to us!!" In other words, if gravity did not propel the water, then it must be pumped. If pumped, then Mars has technology. If technology, then there is intelligence on the planet—a Martian civilization.

The curtain dropped, the applause swelled, and Lowell's mythopoeic unconscious took an appreciative bow.

...

GAINING ACCESS TO THE interior of anyone's mind, especially the mind of a person no longer living, is never easy. In the case of Percival Lowell, I found it a special challenge. Lowell's writing is marked by a stiff formality. Even in letters to friends, he rarely revealed feelings of vulnerability. Although he leavened his paragraphs with puns and bons mots, the overall sense one gets is of a guarded self-consciousness.

My first impression of Lowell was of a man effortlessly clever and confidently in control, but over time I came to see it as an act, an attempt to conceal a bedrock of insecurity. He lived under the weight of expectations passed down from his accomplished ancestors and his demanding father. To even acknowledge those expectations as a burden was to admit weakness and self-doubt—to reveal oneself as a failure—and so he hid those feelings, perhaps even from himself. I was surprised, therefore, to find some hints of emotional truth when reading his correspondence—love letters—from the latter half of 1903.

I had, by this point in my research, uncovered occasional evidence of Lowell's relations with women, which seemed to follow the tropes of powerful men of his era. After he broke off his engagement to Rose Lee and sailed to Asia, back in 1883, he flirted for a time with a Japanese geisha and then apparently took things further with a young woman of similar occupation in Korea. Lowell mentioned her several times in his book about his travels there, and after its publication, the American chargé d'affaires in Seoul wrote a gossipy letter home: "Lowell ought to know something about the women [in Korea] as he lived with the fair creature he describes, and is said to have a child here."

Years later, when Lowell established his observatory in Arizona, he maintained a close relationship with his longtime secretary, who often shuttled between Boston and Flagstaff with him. The two were evidently intimate. Lowell's assistant, A. E. Douglass, discovered as much shortly before he was fired, a fact that I discovered when looking through his unpublished papers at the University of Arizona. One day in 1901, when he went to call on Lowell at his lodgings, Douglass was shocked to find "the young woman's clothing draped over Professor Lowell's trunk." But these were relationships with women over whom Lowell held obvious power. I do not imagine that he felt himself emotionally exposed when he commenced these affairs.

Lowell began a very different relationship in 1903. Having by then reached his late forties, he was ready to find a permanent partner. "When will her Ladyship be at home . . . to a Martian?" he wrote playfully to a woman he had met late that summer at a fashionable beachfront resort in York, Maine. He first spied her at tea, and when she pretended not to notice, he approached and asked for an introduction. Her name was Irva Struthers. She was thirty years old, well bred, well traveled, artistic, and hailed from Philadelphia's social elite. She seemed suitable for matrimony.

With the so-called chivalry of his class, Lowell wooed her. In Maine, he took her for a carriage ride by starlight and shared his thoughts on other worlds. In Boston, he escorted her to the public library to gaze at King Arthur's gallant knights on their quest for the Grail. By the time he visited her in Pennsylvania that December, he was lavishing her with roses and poetry ("To the missed of Earth / The mist of heaven") and was ready to propose. In this, as in all things, Lowell was impatient. He asked Struthers for a prompt answer and a quick wedding—in two weeks—but she did not see how their union could work, for she was devout, he a nonbeliever. She said no. Lowell pleaded with her, in a passive-aggressive manner. "Alone! yes, always alone, though I had hoped so for the opposite. Why does she not save me from it," he agonized in a letter to her. The following year, Struthers accepted the hand of another man. "And so I go back again to my only solace," Lowell wrote, meaning his work. "Perhaps that I may make a success."

The rejection surely undercut Lowell's self-esteem. It also intensified his focus on the orb in his telescope, for Mars gave his life purpose; it offered him the means to prove himself a success worthy of the Lowell pedigree. Like the crazed messenger who stunned parishioners at New York's Calvary Episcopal Church, and like Hélène Smith, that speaker of alien tongue, Lowell was unconsciously shifting his affections, his very identity, toward the people of that faraway world.

CHAPTER
12

Truth in the Negative

1904–1905

As Percival Lowell's fiftieth birthday neared, he drafted a note to Camille Flammarion. "An amusing fact occurred to me that I think you will find a funny coincidence," he wrote from Flagstaff to Paris. "I just learned that I was born half-Martian." Lowell had discovered that the date of his arrival in the world—March 13, 1855—had fallen on a Tuesday. In French, the month of March is *Mars*, and Tuesday is the day of Mars, *Mardi*. "So I am born of Mars not only relative to the month but relative to the day too." It was a French joke—the pun of an American sophisticate. It was also evidence of Lowell's deepening obsession.

By now, Lowell was tackling his work with the fevered energy of a lover spurned. He wrote. He published. He lectured. He lobbied. In the summer of 1904, he sailed for Europe. Arriving in Milan and finding Giovanni Schiaparelli away, Lowell tracked the Italian master to his summer villa, in the rolling countryside toward Lake Como. Lowell ventured out by tram and foot, carrying his latest drawings of the canals. Schiaparelli's eyesight had now deteriorated to the point where he could not study Mars directly through the telescope, but he showed keen interest in Lowell's maps, and also in his graphs—the cartouches—which offered further evidence that the lines on Mars represented plant life that flourished along irrigation channels. "Your theory of vegetation becomes more and more probable," Schiaparelli observed.

From Italy, Lowell made a leisurely traverse of Switzerland, visiting Lucerne and St. Moritz before resuming his work in France. He took the train south from Paris to Juvisy, followed by a carriage ride uphill to Flammarion's observatory.

CAMILLE FLAMMARION AND PERCIVAL LOWELL AT JUVISY.

The French astronomer had invited Lowell to dine with an assemblage of local intellectuals. The men talked of Mars, which was once again approaching Earth. "It was an agreeable table, kind and clever," Lowell noted in his journal. When the meal concluded, Flammarion pulled his Martian friend aside and offered both hands. "Let us hear from you during the opposition," Flammarion implored. He, like the rest of humanity, craved to see what Lowell would achieve at this next close encounter with Mars.

...

LOWELL'S WORK DURING THE previous approach of Mars, in 1903, had heightened the intrigue. "The latest observations of Percival Lowell, who has made himself the special student of Mars before all others, afford us a glimpse of the possibilities of that strange planet as fascinating as a dip into the 'Arabian Nights,'" wrote Garrett P. Serviss with typical flair. "That [Martian] inhabitants actually exist is, of course, simply an inference from suggestive appearances. But these appearances are so extraordinary that no thoughtful observer can very well help trying to interpret their meaning." This enthusiasm was to be expected from the science columnist for the Hearst papers, but it was not just the yellow press that eagerly anticipated what Lowell might find when Mars swung around again, in 1905. The no-longer-condescending *New York Times* now considered Lowell's work, if not convincing, at least fit to print. "We wish to publish your despatches," the paper's managing editor, Carr Van Anda, wrote Lowell. "What we wish for *The Times*," he continued, "is anything new you may detect in your observations of Mars, with the deductions you may draw therefrom."

In scientific circles, those deductions continued to rankle. At England's Greenwich Observatory, Walter Maunder pressed his "small boy theory" that the canals were mere illusion. He chided the irrigation theory as "an excursion into fairyland" and noted a potential way to resolve the debate. Lowell's arguments were based on drawings of Mars, which were inevitably subjective and open to interpretation. What was needed were objective representations of the planet's surface—in other words, photographs. This was, however, a purely hypothetical suggestion. "Mars, unfortunately, does not lend itself to photography," Maunder acknowledged.

Cameras were by this time growing ubiquitous both in daily life and in scientific research. The Eastman Kodak Company had recently introduced the one-dollar Brownie, which made taking snapshots as easy as pressing a button and shipping the entire unit—film and all—to Rochester, New York, for processing. Astronomers, meanwhile, pointed more sophisticated cameras toward the sky to reveal hidden brushstrokes on heaven's canvas. The unblinking gaze of a photographic plate, by collecting feeble light over minutes or hours, could make visible the most dim and diffuse features in space: the braided swirls of spiral nebulae, the wispy tails of comets, the tendrils of the Milky Way. Photography was less successful, however, at recording the faces of planets. "No photograph yet taken of Mars gave more than a small smudged disk," Maunder, at Greenwich, was paraphrased explaining. The reasons were multiple. Mars was tiny and faint, emulsions were slow, and the planet turned on its axis, blurring out delicate markings one might hope to record during a long exposure. Another problem was the *apparent* motion of Mars caused by turbulence in the Earth's atmosphere, which made the image jiggle in the telescope. If one could only tell Mars, as Maunder quipped, "Be motionless, I beg you!"

Lowell had at one time agreed that photographing the details of the Martian surface was "impossible," but now he saw that this seemingly hopeless feat, if it could be achieved, offered his best chance to convince the doubters. He tasked one of his assistants, Carl Otto Lampland, to take on the challenge. Lampland experimented with color filters and sensitive emulsions to shorten exposure times, and he devised a clever way to calm the tremors of jittery Mars. It was a mechanism that, like a movie camera, could capture consecutive images in relatively quick succession. The idea was that if you took enough pictures, you should occasionally get lucky; once in a while, when you opened the shutter—if the atmosphere was still—you just might catch Mars sitting quietly for its portrait.

As the planet's close approach itself approached, Lowell grew anxious. "How gets on the photographing ... of Mars for next opposition?" he wrote Lampland from Boston. "We *must* secure some canals to confound the sceptics."

...

AMERICA'S EARTHBOUND EYES AT this time were focused toward St. Louis. The city had just opened what was the largest international exhibition yet staged, the Louisiana Purchase Exposition, meant to

WORLD'S FAIR, ST. LOUIS, 1904.

commemorate the nation's great westward expansion a century earlier. Better known simply as the St. Louis World's Fair, it was a grand showcase of the country's might and inventiveness, and also its racism and imperialism. Many displays featured so-called primitive peoples and cultures that had now fallen under the control of Western powers. Among the most notable individuals on exhibit, being paid one hundred dollars per month plus income from the sale of souvenirs, was Geronimo, the conquered Apache leader from Arizona.

Millions of fairgoers came: to ride the giant Ferris wheel, to admire the Beaux Arts palaces trimmed with dazzling electric lights, to stroll the midway to the strains of ragtime. The fair also hosted the Olympic Games, America's first, where Garrett P. Serviss Jr.—son of the science journalist and an athlete at Cornell—placed second in the running high jump. Dignitaries and orators came to speak, including a woman for whom that very act seemed unimaginable. "Helen Keller's address at the Hall of Congresses was perhaps the most talked-of speech that has been yet delivered on the grounds," *The St. Louis Republic* reported. "Think of a girl who is totally deaf, blind and who has acquired language, without even having heard it." Keller's words, uttered in a low voice, were repeated by Anne Sullivan, the teacher who had rescued her protégée from utter isolation. Sullivan had famously reached the girl's consciousness across a void of darkness and silence.

"One may ask how we can possibly make the inhabitants of the planets understand us or how we can understand them," a commentator noted a short time later. "This question may be answered with another. How was communication with Helen Keller's mind effected, and perfected until the deaf and blind girl's mind was made to see and hear even though the physical senses remained dormant? If Tesla will furnish the means of communication[,] earthly and Martian ingenuity will devise ways of exchanging ideas."

...

IN MID-DECEMBER 1904, AFTER the world's fair had closed and broad flakes of snow were settling upon New York, Nikola Tesla appeared in elegant evening attire in the mosaic-tiled lobby of the Waldorf-Astoria.

A journalist approached and found the inventor in a talkative mood, ready at last to discuss his mystery tower on Long Island.

Tesla revealed that its immediate purpose was not, in fact, to communicate with the Martians but rather to connect humanity with itself. "It will convert the entire earth into a huge brain, capable of responding in every one of its parts," he said. "By the employment of a number of [towers], each of which can transmit signals to all parts of the world, the news of the globe will be flashed to all points. A cheap and simple receiving device, which might be carried in one's pocket, can be set up anywhere on sea or land, and it will record the world's news as it occurs, or take such special messages as are intended for it." In broad concept, he was imagining a network of smartphones.

His further goals were yet more futuristic. The technology, Tesla said, could also transmit power—vast amounts over great distances—with no need for wires. "We could light houses all over the country," he explained. "We could keep the clocks of the United States going and give every one exact time; we could turn factories, machine shops and mills, small or large, anywhere, and I believe could also navigate the air" by means of wireless energy. It was an astonishing claim, one that rival electricians dismissed as the delusions of a hollow romantic. Tesla fired back with messianic anger and zeal. "My first 'world telegraphy' plant . . . [will strike] the universe with blows—blows that will wake from their slumber the sleepiest electricians, if there be any, on Venus or Mars!" he wrote as 1905 began. "It is not a dream, it is a *simple feat of scientific electrical engineering*, only expensive—blind, faint-hearted, doubting world!"

Lost in the bombast—a mere offhanded remark—was Tesla's mention of the expense, yet this would prove his endeavor's undoing. To construct his great tower, Tesla had secured help from the legendary investment banker J. P. Morgan, a Wall Street colossus who brazenly reorganized the railroads and consolidated the steel companies. Morgan had quietly backed Tesla's wireless scheme after the inventor pitched it to him as a communications technology that could compete with Marconi's. Tesla, however, quickly burned through Morgan's cash, and when he asked for more, Morgan refused. Desperate, Tesla laid off workers, sold his Colorado laboratory for scrap, and begged, cajoled, and shamed Morgan in an effort to pry loose more money. "My enemies have been so successful in representing me as a poet and visionary, that it is absolutely imperative for me to put out something commercial without delay." Yet the banker would not budge.

Tesla seethed, "You are a big man, but your work is wrought in passing form, mine is immortal." Holed up in his rooms at the Waldorf-Astoria, surrounded by luxury yet swallowed by debt, Tesla descended into his own romance of the subliminal imagination. His unconscious mind, like Lowell's, wove a story of personal aggrandizement. "Let me tell you once more. I have perfected the greatest invention of all times," Tesla scolded Morgan. "It is the long sought stone of the philosophers. I need but to complete the plant I have constructed and in one bound humanity will advance centuries." Tesla's monomania and megalomania took over. "I am the *only man* on this earth *to-day*, who has the peculiar knowledge and ability to achieve this wonder and another one may not come in a hundred years."

Tesla's delicate emotional state may have been exacerbated by romantic disappointment; a dashing Spanish–American War hero with whom he had enjoyed an intimate and possibly physical relationship soon announced that he was to marry. "Sometimes we feel so lonely," Tesla once lamented about his own bachelorhood. Increasingly isolated, his finances drained, and his dream tower on Long Island unfinished, Tesla wrote a friend on the letterhead of his palatial hotel. At the bottom of the page, Tesla signed his name: "Nikola Busted."

. . .

FOR LOWELL, WHO SELF-FUNDED his research, money was no constraint, and in the winter of 1905 he returned to his private observatory as the red planet neared. "Canals already seen and in agreement with theory," he wired from Arizona. His telegram soon arrived at the Massachusetts home of Edward S. Morse, the globetrotting zoologist who had long ago inspired Lowell's travels to Japan. Lowell now invited Morse on a journey to Mars, via telescope. He encouraged his friend to come to Flagstaff for an extended stay, long enough to watch the seasonal darkening of the canals and to witness their gemination, "for the doubles do not come on till the early part of June and to miss them would be to see the play minus Hamlet." Morse arrived that spring.

These two men, talkative and opinionated, savored each other's company. They shared meals and cigars and tramps through the pines on Mars Hill. They explored ancient cliff dwellings and hiked through geologic time as they descended switchbacks into the Grand Canyon, all the while marveling at the tenacity of life in this arid environment. Looking across the sublime bleakness of Arizona's Painted Desert, its eroding mesas and buttes striped with mineral pigments, Lowell noted "the resemblance of its lambent saffron to the telescopic tints of the Martian globe." The sight reinforced his compassion for the Earth's extraterrestrial neighbors. "Pitiless as our deserts are, they are but faint forecasts of the state of things existent on Mars at the present time." Morse noted life on a much smaller scale. No matter where he walked in Arizona, even at high altitudes, he found industrious creatures that lived in organized societies and built their own infrastructure: a network of trails and tunnels. "The ant stands among the invertebrates much as man does among the vertebrates," he wrote in admiration of the insects. He praised their "unique intelligence."

At night, the men turned their gaze upward. Morse's initial views of Mars through Lowell's telescope proved inauspicious. "Am in despair of seeing anything," he grumbled, unable to discern the planet's surface features. "I must have an old and worn-out retina." Even three weeks into his observing, he still could not make out the canals. "I . . . have come to the conclusion that it will take months of continuous observation before I can see anything." The following night, however, brought a breakthrough. As if he had wished them into existence, those mysterious and evanescent lines began to appear.

Lowell, with his experienced eyes and biased mind, saw much more on Mars, and he wired details of his observations to the Harvard College Observatory, which served as a clearinghouse for astronomical news. He reported new evidence that the Martian farmers were cooperatively distributing water: "Enigmatic functional alternation in visibility of Thoth and Amenthes on Mars has again been observed

EDWARD S. MORSE AT LOWELL'S TELESCOPE.

here." He announced an obvious change in the seasons: "First winter snow ... in arctic region of Mars." Then a small item on the front of *The New York Times* brought the biggest news of all:

> CAMBRIDGE, Mass., May 27.—A telegram was received at the Harvard Observatory to-night from Prof. Percival Lowell, Director of the Lowell Observatory at Flagstaff, Ariz., stating that the canals of Mars have been photographed there for the first time.

The New York *Sun* soon quoted Lowell more forcefully: "To-day we can state as positive and final that there are canals on Mars—because the photographs say so, and a photographic negative is nothing if not truthful."

...

ONE CAN FIND THOSE photos today, originals, in the archives of the Lowell Observatory. They consist of monochrome contact sheets—

some black-and-white, others in shades of brown—that depict column after column of shrunken globes, like rolling agates in a game of marbles. In these pictures, Mars is minuscule, the planet's disk less than a quarter inch across ("hardly larger than the head of a shirt-stud," a commentator in 1905 had noted), the small size having been necessary to shorten the exposure and enlargement impossible without making the prints unacceptably coarse and grainy.

On one of my visits to Flagstaff, I held the tiny photos up to my face and squinted. Here was Lowell's supposed proof, but what exactly did the pictures show? I could see that the planet was marked with dark patches in various shapes—one looked like a *y*, another like the Greek letter lambda (λ)—yet I jotted in a note to myself, "Where are the canals?" To this layperson, and to many in 1905, the Mars photographs were indecipherable inkblots. It was therefore left to experts to decode them.

"Here they are," Flammarion said in October 1905 as he emptied an envelope at his observatory, in Juvisy. Out spilled seven tiny Mars prints sent by Lowell. "They are remarkably sharp." The French astronomer showed the photos to a visiting journalist and pointed excitedly to a patch of white that appeared toward the bottom of each image: "The polar cap!"

PHOTOGRAPH OF MARS BY C. O. LAMPLAND.

The reporter was more interested, however, in any evidence of artificial constructions. "So, Monsieur Flammarion, do you firmly believe in the existence of the Martians?"

"Yes, I believe it!" the astronomer proclaimed. He gestured again toward the photos. "Examined with a good magnifying glass, they show us the following canals," he continued, naming eight Martian waterways that he recognized.

Other astronomers in Europe and the United States were less certain about what the pictures revealed. Sir Robert Ball, that avowed canalist, confessed that he saw no lines: "To make them out would have required better eyes than I have got." Oxford's H. H. Turner, though genial toward Lowell, also sat on the fence: "Personally I find it extremely difficult to say exactly what is on those prints, and friends to whom I have shown them differ a good deal in their interpretations." Most favorably for Lowell, the head of the British Astronomical Association—the group that had laughed at Lowell when Maunder, from Greenwich, presented his "small boy theory"—examined the pictures and found himself swayed. "The canals were clear and unmistakable, appearing as continuous narrow, slightly curved lines. These photographs did a great deal to strengthen my faith in the objective reality of the canals."

Despite the confusion over what the photographs showed, they were undoubtedly the best yet taken of Mars, a technological coup that augured well for the future. In two years, the planet would come around again, and this time it would sail closer, shine more brightly, and therefore be easier to capture in a portrait.

For now, Mars once again receded and dimmed amid the stars in the night sky, and Lowell prepared to return to Boston. Before leaving Flagstaff, he gathered another batch of photographs to post in the mail. These pictures were terrestrial: snapshots of Edward Morse peeping through the telescope, hiking the Grand Canyon, relaxing at the Baronial Mansion. Lowell slipped the photographs into an envelope and addressed it to the friend with whom he had traveled both Arizona and the solar system. He composed an accompanying note. It opened, "Dear fellow-Martian . . ."

CHAPTER

13

Planet of Peace

1906

At the 1904 world's fair in St. Louis, among the most popular exhibits had been those that stood at the muddy western end of the expansive fairgrounds. These were the Anthropology Villages. "Plans have been completed for assembling . . . representatives of all the world's races, ranging from smallest pygmies to the most gigantic peoples, from the darkest blacks to the dominant whites," boasted the man in charge, one WJ McGee, wielding the noxious language

AFRICAN VILLAGE, ST. LOUIS WORLD'S FAIR.

typical of the time. McGee was a prominent ethnologist in his day, although history has not remembered him well. ("McGee was an inexhaustible mine of every error of substance and theory that it was possible to commit on the basis of the most vulgar prejudices," one modern anthropologist has written.) Embracing the racist assumption of Anglo-Saxon superiority, McGee had, for the St. Louis World's Fair, collected Indigenous peoples from across the globe like so many natural history specimens. Enticing them with money and gifts, he placed them like props in a living diorama.

Hunter-gatherers from Central Africa posed half-naked before huts woven from sticks and leaves. Nearby, where human skulls adorned a bamboo fence, loinclothed tribesmen from the Philippines feasted on dogmeat. Fairgoers toting Kodaks insolently crowded the exhibits as if eyeballing animals in a zoo. "The Americans treat us as they do our pet monkey," complained one of those on display in the African village. "They laugh at us and poke their umbrellas into our faces." McGee, the anthropologist, protested the rude treatment of his Indigenous guests, yet he well understood—indeed *exploited*—what he called "that innate and intuitive curiosity which renders alien races so attractive to all mankind."

This curiosity about alien races spilled onto the planet next door in early 1906 when the Hearst papers, in their Sunday editions, ran an artist's rendition of possible Martian life-forms. Here for millions of readers to ogle was a motley assemblage, a veritable Anthropology Village from Mars, complete with exaggerated eyes and lips that disquietingly evoked the racist imagery of blackface. Of course, Martians were not human beings, yet they were often imagined that way—on stage, in books, in séances. "I cannot understand why people persist in picturing the inhabitants of the planets as men," the Oxford astronomer H. H. Turner had recently complained. "It should be borne in mind that we are carefully adapted to the conditions of life on the earth, and as the conditions vary on the planets so do the forms of life vary, so as to be adapted to these conditions." In truth, the new depictions of Martian life embraced this somewhat more scientific attitude. The illustrations showed, based on the spec-

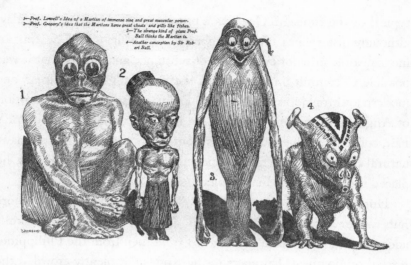

HOW VARIOUS MODERN SCIENTISTS HAVE FIGURED OUT THAT A DWELLER ON THE PLANET MARS WOULD PROBABLY LOOK.

ulation of various experts, how alien anatomy may have been molded by the unique conditions of Mars.

Take, for instance, the force exerted by Mars itself, which scientists could easily calculate based on the planet's known mass and size. "Gravity on the surface of Mars is only a little more than one third what it is on the surface of the Earth," noted Percival Lowell, who explained that this low-gravity environment should make the Martians appear especially strong. "To begin with, three times as much work, as for example, in digging a canal, could be done by the same expenditure of muscular force." Moreover, the Martians might actually *be* stronger, and bigger, than humans. Gravity constrains how massive an animal may grow—if a creature becomes too big and heavy, its framework of bone and muscle will no longer support it—so under conditions of reduced gravity on a small planet like Mars, evolution might build animals on a larger scale. Sir Robert Ball noted this seeming paradox. "At first it might be supposed that big animals might be most appropriately located on big worlds, and small animals on small worlds," he wrote, yet "the truth lies the other way."

Another factor to consider was the planet's scant air. This condition was addressed by Richard Gregory, a friend of H. G. Wells and an editor at the British journal *Nature*, which for many years gave Lowell's theories favorable coverage. "Whatever atmosphere exists on Mars must be much thinner than ours, and far too rare to sustain the life of a people with our limited lung capacity," Gregory wrote. He then surmised an evolutionary adaptation: "A race with immense chests could live under such conditions." Other thinkers inferred a second adaptation to the weak atmosphere. Since the dilute air would transmit smells poorly, one wrote, "the man on Mars [may] have an elephantine nose," like a trunk, "which goes to the odor, as the odor can't come to the Martian."

As for their sense of sight, Martians should have "very big eyes—wonderful eyes—according to scientific theory," explained Arthur Brisbane, the editor who long sensationalized Mars at both the Hearst and Pulitzer papers. "Mars is so far from the sun that the big eyes would be required to gather in as much of the sunlight as possible." Brisbane further asserted, on somewhat looser logic, that those eyes should be set in a large and egg-shaped head. He argued that human beings had evolved large brains and lost their body hair over the generations (because, presumably, we descended from an apelike human ancestor hypothesized by Darwin), and Martians had progressed further down that same developmental path. "As Mars is millions of years older than we are it is very probable that the heads of the Mars men are something tremendous and that they constitute a large part of the entire man," Brisbane wrote. "It is also likely that the man and woman of Mars are absolutely bald. This seems at first a repulsive state of things, but it should not appear so. Doubtless, where we have hair they have rosy and attractive scalps."

These traits—towering body, broad chest, gaping eyes, bulbous and bald head—characterized the new depictions of Martians, yet still the hypothetical creatures hewed close to human form; they possessed two hands, two feet, a backbone. A modification to this body structure soon appeared when Lowell's fellow-Martian Edward S. Morse, after returning to Massachusetts from his stay in Arizona, wrote a book that defended Lowell's canal theory from his perspective as a zoologist. Morse noted that some intelligent species on Earth live in dry, high-altitude environments and might, after countless generations of evolutionary change, adapt to the even drier, more rarefied atmospheric conditions of Mars. In other words, Mars might be inhabited by "precisely the kinds [of creatures] that thrive under the most diverse conditions here—namely, man and . . . the ants." That statement spawned sensational headlines—PROF. E. S. MORSE THINKS [THE MARTIANS] MAY BE GIANT ANTS, the Boston *Sunday Herald* declared—implanting a new idea in the public mind. Some illustrators now added antennae to their alien depictions, and thus did the Martians evolve.

MARS PEOPLE AS THEY MAY LOOK.

. . .

THE EARTH SEEMED TO be spinning in Lowell's direction. "Since many of the 'canals' were photographed last Spring," Garrett P. Serviss noted, "there has not been so much laughter at the expense of the Flagstaff astronomers." The papers offered gushing profiles of the aristocrat "recognized throughout the world as an authority on the solar system," an astronomer "whose logical deductions [are] as full of fascination as Jules Verne's romancing." As for that other famous author of speculative fiction, the British writer H. G. Wells, modern scholars have puzzled over whether he and Lowell ever met, because neither man left a record in his correspondence or writings.

As I discovered when combing through newspapers from the era, however, they did meet around this very time.

In mid-April 1906, Wells—who was by then celebrated not only as a novelist but as an incisive social commentator—made his first trip to the United States. As he steamed into New York Harbor, he found himself awed by the skyscrapers that dwarfed his ocean liner, and over the next month he continued to take the measure of this young and ambitious country. He visited President Theodore Roosevelt at the White House and the social reformer Jane Addams at Chicago's Hull House, but his first stop after New York was New England. "He is soon to visit Boston, where he will meet several acquaintances and literary men," *The Boston Post* reported. Indeed, on Monday, April 23, still bronzed from his transatlantic voyage, Wells attended an afternoon reception with prominent Boston authors. It is almost certain that at this gathering Lowell and Wells, the two men of Earth so closely aligned with Mars, finally conversed.

Wells mentioned the meeting the following month as he prepared to leave New York for his passage home. A reporter, after asking the British writer his impressions of America, inquired if he believed that interplanetary communication was possible. "Why not?" Wells replied. "In a talk with a noted scientist in Boston I was shown things and told things about Mars which were astounding to the imagination. It seemed to place the habitable condition of that planet almost beyond a doubt." Lest there be any question as to the identity of that "noted scientist," Wells remarked to another journalist, "Prof. Lowell told me many things that were simply amazing."

Further evidence of Lowell's rebounding reputation soon emerged from the American Southwest. In Arizona that summer of 1906, the talk was of statehood, which Congress had provisionally granted the territory subject to a controversial requirement: that its citizens agree to a merger with New Mexico. In a letter to an Albuquerque newspaper, a writer from Flagstaff suggested a way to sweeten the deal for Arizonans: "You could give us the first governor—say Percival Lowell, one of the best men [in] every way in the southwest." No such arrangement was struck, however, and the people of Arizona, unwilling to share power with the people of New Mexico, rejected joint statehood.

New Yorkers, meanwhile, were divided by a notorious controversy over race. One of the Indigenous tribesmen who had originally come to America for the St. Louis World's Fair was now, due to an unsettling sequence of events, living at the Bronx Zoo, where he had been brazenly placed in the primate house, behind a sign that read THE AFRICAN PIGMY. The outlandish display drew crowds of shameless visitors and howls of justifiable protest. "We do not like this exhibition of one of our race with the monkeys," fumed a committee of Black Baptist ministers that eventually secured the man's release. "We think we are worthy of being considered human beings, with souls."

Such discord was typical of earthlings, who divided themselves by politics, by blood, by skin, by tribe. Lowell's Martians, in contrast, worked as one. "The first thing that is forced on us in conclusion is the necessarily intelligent and non-bellicose character of the community," Lowell now wrote in his latest book, a sweeping treatise called *Mars and Its Canals* that Lowell's mentor, Schiaparelli, praised as "quite simply a small masterpiece." The book reiterated many of Lowell's previously established astronomical ideas but bolstered them with additional evidence cherry-picked from other fields—geology, biology, paleontology—that gave the theory a feeling of new heft and credibility. In the final chapter, Lowell went so far as to extrapolate about Martian culture: "When a planet has attained to the age of advancing decrepitude, and the remnant of its water supply resides simply in its polar caps, these can only be effectively tapped for the benefit of the inhabitants when arctic and equatorial peoples are at one." In other words, the Martians had learned out of environmental necessity to think and act as a global whole. "Difference of policy on the question of the all-important water supply means nothing short of death."

What Lowell presented was a compelling image, one that inspired others to dream. "[Mars] is like a sinking ship, able to stay afloat only as long as its 'people' work desperately at the pumps—the pumps being in this case the canals," the *Boston Evening Transcript* told its readers.

"It is self-evident [the planet] has no opposing nations and countries, wars and rumors of wars, diplomatics and politics; for the canals all work in perfect unison, and are plainly the result of an ideal coöperation for the common good. So instead of being the Planet of War he should be rechristened the Planet of Peace."

...

BY MID-OCTOBER 1906, THE maples in Boston had turned and the evening was brisk when a procession of carriages and motorcars filled Copley Square as if it were grand opera night. The cultured crowd descended from their vehicles and climbed the stairs of MIT's Rogers Building, that stately edifice of granite and brick, where the throng quickly overflowed the auditorium. On this night, the hall's nine hundred seats and standing room proved inadequate to meet the demand.

The great attraction was the first lecture of the new season of the Lowell Institute, the opening of a course titled "Mars as the Abode of Life." With many attendees excluded for lack of space, the institute's curator stepped forward to appease those disappointed; he announced that the speaker, Percival Lowell, had kindly consented to return and repeat his lecture to accommodate those turned away. When the doors finally closed, onto the stage emerged the dapper Martian himself.

It was in this same hall that Lowell had spoken about Mars to memorable effect in 1895. Since that time he had "not been idle,"

PROFESSOR PERCIVAL LOWELL.

Lowell now assured the audience, and he promised that this new round of lectures would offer the fruit of a decade's observation and thought. Over the next month, on two evenings each week—and also two afternoons, for the overflow crowds—he slowly unspooled his tale with his usual earnestness and charm. "Mars is a vast Saharan world, where fertile spots are the exception, not the rule, and where water is everywhere scarce. It is a saddening picture of a world in destitution, a world athirst, where water is the one thing needful, and yet where only from the annual melting of the polar snows can any of it be got."

He reviewed his latest evidence: the cartouches, the wave of darkening, C. O. Lampland's tiny photographs. Although Lowell would not speculate whether the Martians were humanlike, antlike, or something entirely novel, he found their existence and peaceful nature undeniable. "Not only do the observations lead us to the conclusion that Mars is at this moment inhabited, but they land us at the further one that these denizens are of an order whose acquaintanceship is worth the making." He flashed a playful smile.

Not all in the audience fell under his spell. "Some questioned the facts set forth; others, the methods of reasoning," reported a journalist friend of Lowell's. One woman leaving the hall was overheard to say, "Well, I call him a visionary!"—by which she meant a dreamer, the same epithet so often hurled at Tesla. Yet the overall response was decidedly enthusiastic, and the excitement rippled outward. "In many a household," *The Boston Daily Globe* noted, "Mars receives more attention than ever before, on account of the treatment accorded by Prof. Lowell in these lectures."

Among the fans in the audience was that sunny, prolific writer Lilian Whiting, who in addition to penning books and columns, attending social events, and traveling the world, was now contributing features to *The New York Times*. She offered the paper a lengthy article about "the great astronomer who is now held to be the specialist on Mars—the Martian expert, as it were." Editors at the buttoned-down *Times* not only accepted the piece but published it prominently in the Sunday edition and topped it with a headline both bold and definitive: THERE IS LIFE ON THE PLANET MARS.

In this case, however, *The New York Herald* offered a more nuanced headline. The coming year would provide the best opportunity to study Mars in fifteen years, with the planet coming

closer to Earth than at any time since the "boom" of 1892, and the canalists hoped that any lingering doubts about life on Mars might soon be settled. So, on the final Sunday of 1906, when the *Herald* published a feature about Lowell that quoted liberally from his latest book, the paper topped the article not with a statement, but a question: WILL THE NEW YEAR SOLVE THE RIDDLE OF MARS?

CHAPTER
14

Alianza

1907

Percival Lowell, through his telescope, often pondered the polar regions on Mars, for he believed they were the source of the planet's life-giving waters. In 1905, when he had seen and reported the first snows of winter in the Martian Arctic, the public whimsically imagined that distant, frozen scene. "[We] can see [the] lively and intelligent little inhabitants donning their overcoats and winter wraps, storing up their winter fuel, and gliding over the crisp snow to the tune of merry sleigh-bells," *The Los Angeles Times* had mused, anthropomorphizing the aliens as was often done. The Washington *Evening Star* had injected at least a hint of objective science into its own humor when it expressed sympathy for the "unlucky Martians" with benumbed extremities and frostbitten ears and noses—"if they have heads and feet and ears and noses," the paper clarified.

Now in February 1907, a blizzard pummeled Boston. The icy gale halted locomotives in their tracks, shuttered the schools and the courts, and whipped up towering drifts that turned the roads into canyons. Lowell ventured into the storm, pushing through the heart of the city as flakes collected on his shoulders. "This is a real taste of winter," exclaimed a passerby who recognized the famous astronomer.

"Yes, indeed," Lowell replied with a smile. "It is very Martian in its severity."

No longer the frail aristocrat who had once suffered nervous exhaustion, Lowell exuded confidence and optimism, for he had just launched a new scheme—his most ambitious yet—to silence his critics.

On a night the previous November when another frigid wind chilled Boston, a train from the north—the "Flying Yankee"—chuffed into the city draped with icicles. A crowd on the platform surged forward as a tall figure with a feathered mustache emerged from the Pullman car. This was Commander Robert E. Peary, who moved through the throng with a quick and confident step despite having lost his toes to frostbite. Reporters shouted and news cameras flashed. "No one who is interested in the wonderful development of the twentieth century can escape being fascinated by the outlook of [Commander] Peary for reaching the North Pole," Lilian Whiting raved about this American hero who lusted for immortal fame.

Peary, like Lowell, was indefatigable—obstacles seemed only to spur his obsessiveness—and he was returning from yet another trip to the Arctic, a region he called "that gaunt frozen border land which lies between God's countries and inter-stellar space." With so much of the world now trod and mapped, the extreme top and bottom of the Earth had become the coveted prizes of exploration. "There is no higher, purer field of international rivalry than the struggle for the north pole," said Peary, who had been striving for a decade to plant the American flag there. Although his efforts had once again failed, he had succeeded this time in setting a new "farthest north" record, reaching a point less than 3 degrees shy of his goal (or so he claimed). For this feat, President Roosevelt would bestow on Peary a gold medal from the National Geographic Society, a young organization that promoted science and global exploration. At the awards banquet, Peary would in turn praise Roosevelt for his own effort to impose American dominion on the Earth, by connecting the oceans. "Between those two great cosmic boundaries, the Panama Canal and the North Pole," Peary would say, "lie the heritage and the stupendous future of . . . the United States of America."

During his brief stopover in Boston in November 1906, Robert Peary and his wife, Josephine—who had accompanied her husband on previous Arctic adventures and in this case joined him for

ROBERT AND JOSEPHINE PEARY.

the trip home—were whisked by automobile to a private reception with some of the city's elite. Lowell was there, and the two men, these explorers of worlds, conversed at length about matters scientific. Lowell and Peary moved in similar circles—both were members of the exclusive Ends of the Earth Club—and people sometimes spoke of them in the same breath. Lilian Whiting compared Peary's daring trek to "a journey to Mars," and Lowell, a student of the Martian ice caps, considered himself a polar adventurer, albeit on another planet.

Of course, Lowell's journeys by telescope did not require the same physical effort as Peary's. Lowell struggled merely to stay awake as he made nocturnal trips to Mars from the observatory dome beside his Baronial Mansion; he did not brave the jagged ice of the Arctic Ocean, did not live on rations, did not risk losing his toes or his life. This lack of hardship made Lowell's form of exploration "less appealing to the gallery," as he put it, but perhaps emboldened by his meeting with Peary, he soon developed his own plan to enthrall the masses with an arduous earthbound expedition. In his case, however, Lowell would not have to suffer the discomfort. He would pay others to endure the slog.

...

AMERICA'S ECONOMIC MOOD, so buoyant during the early years of Roosevelt's presidency, betrayed a notable souring in the spring of 1907. The nation's financial system had recently experienced several shocks—including a literal one, the San Francisco earthquake of 1906, which leveled the city and rattled the markets, leading to a slump on Wall Street. By May 1907, however, stocks had stabilized and the worst seemed to be over when Americans found a new cause for excitement. "The riddle of the planet Mars may be brought nearer solution by the Lowell expedition, headed by Prof. Todd, of Amherst College," read a widely printed wire story, which announced that the scientific enterprise had just left for South America.

David Todd, the intrepid astronomer who had joined with Lowell in 1900 to observe that year's total solar eclipse in Tripoli, had now launched a new venture for his and Lowell's benefit. As both men well knew, the upcoming opposition of Mars, in July, would bring that planet closer than it had been in a decade and a half. Flagstaff, however, would not be an ideal observation post this time; in fact, nowhere in the Northern Hemisphere was well situated. Mars would be positioned toward the south, which meant the best views of the planet would be had from below the equator, where few large telescopes were permanently stationed.

It so happened that July 1907 would also bring a solar eclipse to South America, and Todd was keen to observe it, so he made a proposal to his wealthy friend. If Lowell would provide the money, Todd would take Amherst College's telescope—a twenty-five-foot, seven-ton monster—on a five-thousand-mile journey to the Andes Mountains. There he would observe the eclipsed sun for himself and take new photographs of Mars for Lowell. These images, he expected, would be clearer than the ones Lowell had acquired in 1905 because the planet would now be closer, brighter, and bigger. One might even hope that the new pictures could capture those most elusive of all Martian features, the double canals.

Lowell assented, offering not only funds for the venture but also his special photographic apparatus and an assistant trained to operate it. Unsurprisingly, Lowell demanded something in return: the naming rights for the enterprise, which he dubbed the Lowell Expedition to the Andes, a brand emblazoned on custom stationery and luggage tags. When Giovanni Schiaparelli heard the news, he sent his congratulations. "I hope that [this campaign] is even more fortunate than all your previous ones, and that's saying a lot." Camille Flammarion also offered encouragement: "How exciting if this year you secure even more complete photographs than the first ones!"

Before departing New York on the SS *Panama*, a gregarious David Todd, always eager to talk to the press, described the expedition's ambitious plans. "This will be the largest telescope ever used in South America and after we have selected some high point in the Andes, where the atmosphere is clear and steady, we will mount our telescope and begin operations." He then

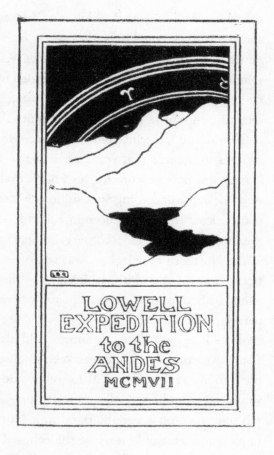

turned philosophical. "'Is Mars inhabited?' That is a question my wife has often asked me, but do you know I have never been able to answer it. Maybe when I come back—well, we will wait and see." As for his wife—the famed writer and lecturer, Mabel Loomis Todd—she, too, was to join this tropical adventure. "She will be the Mrs. Peary of the expedition," one journalist quipped, "only that the climatic drawbacks to be encountered will be excessive heat instead of cold."

On the day of departure, Lowell telegraphed David Todd at the pier: "Don't forget tooth brush and telescope. May be useful." Todd answered dryly, "Curiously enough 'tooth-brush' was forgotten," but the telescope was snug in the cargo hold, its disassembled parts packed in 119 boxes. The two fragile lenses, each as large as a dartboard, were stored separately. To guard against dust and scratches and breakage, the glass disks had been sewn into flannel pouches, wrapped in tissue paper, glued inside manila bags, and cushioned in cork-filled boxes that Mabel Todd guarded closely until she could stow them someplace secure, such as the captain's cabin or the ship's safe. David Todd later recalled that protecting the lenses took as much care as tending twin babies. "Almost the only difference was that they didn't require feeding, . . . and they didn't cry when left alone."

...

ON THE WAY TO South America, the expedition paused for several days in Panama, where Roosevelt's artificial waterway across the isthmus, to become one of the great macro-engineering feats of the twentieth century, was far from complete. Disease outbreaks, mudslides, and grinding bureaucracy had so plagued construction that it almost derailed. As one newspaper teased, reflecting a widely shared sentiment: "Before we dig any farther on that canal it might be well to have Mr. Tesla wire the government of Mars and ascertain just how they did it up there without bankrupting the planet."

By the time the Todds arrived at the Canal Zone, in mid-1907, the project's fortunes had brightened. Infectious mosquitoes were under control, housing had been upgraded, and the general morale and workflow were much improved. Mabel Todd was excited to tour the site, which offered material for future

"Hello, Uncle Sammy! How many thousand miles of canal have you dug this fall?"

lectures. She was most impressed by the visit to Culebra, where a hive of workmen and machines was tearing down mountains to carve an improbable path for ships through the Continental Divide.

"Have seen a lot of this terrene canal," David Todd wrote to Lowell, "and now we go on to those of Mars."

After crossing the isthmus by train, the Todds and their telescope boarded a new ship to continue south. Down the Pacific Coast they steamed, along Ecuador and Peru, looking for a suitable place to land. Persistent clouds and outbreaks of yellow fever at various ports kept them moving until they arrived at Iquique, a small city in northern Chile that hugged the coast against towering dunes. The community was prosperous, as it exported to the world the means of both life and death in the form of nitrates—raw material for fertilizer and explosives—wrung from the Atacama Desert by a British-dominated mining industry.

Disembarking with the telescope and letters of introduction, the Todds arranged logistics, then boarded a railroad that gently zigzagged up the sand cliff and onto an elevated plain. Andean volcanoes loomed in the distance, their tops clad in snow, while the foreground appeared desolate and unearthly. Almost nothing grew in the desert's hardpacked soil, its colors grading from ochre to russet to rust. "A trip to the moon, or even Mars, could hardly be more interesting," David remarked. The train passed alkali flats and the bleached bones of abandoned draft animals, then came to a string of industrial encampments called saltpeter towns or *oficinas*, where laborers mined and processed the nitrates. At the largest of these outposts, Oficina Alianza, the Lowell Expedition and its giant telescope at last found a place to rest.

Alianza operated like a mining colony on the moon, or perhaps a company town in hell. Several thousand dust-covered laborers slept in cramped and fetid barracks of corrugated iron, worked blistering days setting explosives and hauling ore, and lived amid the rumble and smoke from the adjacent processing plant. In this driest of deserts, where food and supplies had to be imported from a great distance, the employees depended for their very existence on company stores—a canteen, a bakery, a market—at which they relinquished their meager wages back to the mine owners. A loaf of bread might cost a peso, a quarter of a day's earnings.

The bosses, meanwhile, lived at a genteel distance, occupying an airy administration building that perched on a rise. The broad, green bungalow was fringed by a veranda, and its interior featured electric lights, a library, a staff of domestics, one kitten, five dogs, and an assortment of birds. ("I have failed, so far, to win the affections of the parrot," Mabel Todd deadpanned.) The opulence, jarring amid such desolation, provided the visiting scientists with more than creature comforts. David Todd required a flat and stable base on which to erect his telescope. When he looked outside, he found the ideal platform: a concrete tennis court.

DAVID AND MABEL TODD WITH THE AMHERST TELESCOPE AT ALIANZA, CHILE.

The mine managers, thrilled for some distraction from the usual drudgery, offered cozy lodgings as well as mules, men, and hoists to help reassemble the great eye from Amherst. Soon, its delicate lenses were back in the tube, and while Mabel Todd toured the surroundings and practiced her Spanish, her husband prepared for his campaign of photography. "Through all these clear and wonderful nights, David is working at Mars, *el planeta Marte*," she wrote in her journal. The planet shone high overhead at midnight, a garnet among the diamonds of Sagittarius.

By early July—two months after leaving New York—David began his work in earnest, and Mabel Todd had an opportunity to gaze through the telescope. She could see the surface of Mars, 38 million miles away, with what seemed to be exceptional clarity. There, staring at her through the instrument, were the planet's white polar caps, a dark oasis, and a half dozen canals. When she pulled away from the glass, her eyes welled with tears. "It is awe-inspiring and mighty to actually see all Mr. Lowell has drawn," she confided in her diary.

...

COMPARED WITH THE VIEW of Mars through the desert skies of Alianza, the perspective from Flagstaff was predictably inferior. The planet sat lower toward the horizon and thus appeared more distorted by Earth's atmosphere. Still, Lowell spent these nights watching the alien world while, during the day, he savored the rustic affluence of the Baronial Mansion. That spring and summer, as we know from his archived business correspondence, he ordered express shipments of French sardines and Roquefort cheese from Los Angeles. From Boston: claret and cigars. Meanwhile, he managed publicity for his expedition. *The New York Times* once again telegraphed requests for updates. "All the world is patiently waiting for news from Mars," the paper told its readers. The yellow dailies were also eager. "Has the secret of life on Mars been wrested from space?" asked Hearst's *Boston American*. "It remains for the results of the Lowell expedition to the Andes to give us the latest facts about our nearest planetary neighbor."

The expectations weighed heavily on Lowell, for he once again faced attacks on his credibility. Over the years, whenever he was asked why so few astronomers had reliably seen the Martian canals when he saw them so often, he offered two stock replies: The viewing conditions in Flagstaff, he claimed, were far better than elsewhere, and no observatory studied the planet more assiduously than his did. Those answers were suddenly undermined, however, by Lowell's first astronomical assistant, Andrew Ellicott Douglass, the man who had watched the planet from Mars Hill for seven years until his letter to Lowell's brother-in-law caused his unceremonious dismissal in 1901. Now a professor at the University of Arizona, Douglass used his new platform to proclaim himself an anti-canalist. Word of this embarrassing defection from Lowell's camp brought glee to England's Greenwich Observatory. "Professor A. E. Douglass, who . . . worked at Flagstaff for a considerable time, has dissociated himself most emphatically from Mr. Lowell's conclusions," Walter Maunder gloated. "It is one of life's little ironies that Mr. Lowell, who has spent much time, pains, and money on the study of Mars, should be known, not for his real work upon the planet, but for the fairy-tale with which he has garnished that work."

Lowell did his best to bury the news. He tarred Douglass as a disgruntled former employee, a man he claimed to have fired "for untrustworthiness." He inoculated himself against further attack with a dram of public adulation when he accepted an honorary degree from Amherst College, an accolade he had negotiated as a quid pro quo for having funded David Todd's expedition to Chile. What Lowell most needed, however, was the same thing the press so dearly sought—sensational news from that desert outpost in remote South America.

. . .

AT ALIANZA, DAVID TODD and his able assistants were by now taking hundreds of pictures of Mars each night. The tiny globes ran up and down the photographic plates, which the men promptly developed in a makeshift darkroom. Before the expedition left the United States, Lowell had asked the team to alert him by cable should the negatives reveal anything of note, and he established a cypher code to shorten messages, reducing their cost and shielding their meaning from the prying eyes of telegraphers. In this secret language, the word SPOT meant "oases photographed." DOUBLE indicated "canals seen double." Other codes referred to specific canals. NIL, for instance, signified "Nilo Syrtis photographed double." FEEBLE meant "canals light in southern region."

On July 4, Mabel Todd awoke at 6:30 A.M. to firecrackers and a bugle, and when she ran outside in her kimono and slippers, she found the mine managers' bungalow decorated with flags and pennants in honor of America's Independence Day. The celebration continued through the afternoon and into the night, when—in the big library—the Todds and their hosts, lubricated by alcohol, sang by the fire and enjoyed the cueca, a South American dance that mimics the courtship behavior of a rooster and hen, handkerchiefs twirling like feathers.

For the astronomers in attendance, there was more to celebrate than the Fourth. That day, David Todd had cabled Flagstaff: NIL SPOT DOUBLE FEEBLE. An elated Percival Lowell translated the message and notified the world. "The results have proved all that could be wished," he wired *The New York Times*. "Cables have been received stating that both canals and oases have already been photographed there."

DANCING THE CUECA.

In the following weeks, the team in Chile sent thrilling news of many more canals photographed, some of them double, including the Thoth and Amenthes, the adjacent irrigation channels whose alternation suggested an amiable sharing of resources by Martian farmers. By the time the telescope was dismantled at the start of August, the expedition had accumulated more than ten thousand photographs of Mars. A giddy David Todd wrote to Lowell: "As they looked at these photographs and then flew over the pampa, even the bats of Alianza screamed *oasis, oasis, oasis—canali, canali, canali!*"

Todd's benefactor could hardly contain his pleasure.

"Bravo!" came Lowell's salute from Flagstaff. "Your despatches cause our hair to stand on end and our voices to stick in our throats." Jubilant, Lowell placed for himself another express order from Los Angeles: one case of Mumm's Extra Dry Champagne.

...

THROUGH LATE SUMMER AND early fall, as the Todds sailed unhurriedly home—including a side trip to Peru and a return visit to

Panama—the American economy sagged. The financial troubles in the spring had spawned a recession, and instability then devolved into crisis.

On a fine October Tuesday in New York, adjacent to the towering palace that was the Waldorf-Astoria, luxury carriages and motorcars lined Fifth Avenue before a marble-columned mausoleum of a building, home to the Knickerbocker Trust Company. A prominent financial institution that served as a kind of bank to the city's merchants and millionaires, it had recently been the subject of troubling rumors, and hundreds of depositors now squeezed inside the cavernous lobby, beautifully detailed with mahogany and brass. The patrons demanded their money, which they briskly stuffed into satchels and valises until the perspiring tellers, out of cash, informed the crowd that they could honor no additional withdrawals. Thus began the Panic of 1907, which would trigger more bank runs, bring the financial system nearly to its knees, and lead, eventually, to the creation of the U.S. Federal Reserve System to prevent future such disasters.

The very day after the run on the Knickerbocker, when *The Wall Street Journal* warned that "the financial markets are in a state of distress," the SS *Colon* sailed into New York bearing the cheerful Todds, David still dressed for warmer climes in a straw hat and ice-cream suit. Despite the financial scare that dominated the news, the Todds, like Commander Peary upon his return from the Arctic, became bait for the press. Reporters cornered them at the dock and chased them down in the city.

"What sort of people live on Mars?" the newsmen wanted to know.

"The camera has not discovered any people, walking or flying, or floating about on that planet so far," David Todd told *The New York Times* with a smile, "but the whole trip was full of surprises."

Mabel Todd talked to *The Sun*: "I suppose the thing most interesting about the expedition is that for the first time the double or parallel canals have 'come up' on a photographic plate. . . . There were many doubting Thomases who maintained that observers were 'seeing double'—that the whole thing was an optical illusion."

David Todd spoke more forcefully in an interview for the Hearst papers. "I have proved," he said, "that Mars has double canals that could only be the work of intelligent beings living there." Then, as if he had not been clear enough, he added, "I have no doubt that Mars is inhabited."

The news spread at the speed of telegraphy. DOUBTS ARE ALL DISPELLED proclaimed a headline in Massachusetts. PEOPLE IN MARS? SURE THERE ARE, asked and answered a paper in Iowa. As far away as Juneau, Alaska, readers caught the news: MARS PROVEN INHABITED. Still, the public had not actually seen the pictures from Chile, and a bidding war ensued among the nation's great periodicals, each vying for the right to print the images.

Lowell remained coy. "I have your telegram and beg to say that the photographs of Mars are not for newspaper reproduction," he wrote *The New York Times*. "They are extremely valuable and there is no known process suitable to a newspaper that can do them justice." After considering his options, Lowell chose to publish his precious photos and an article to go with them in one of America's most storied magazines.

...

DIGNIFIED AND AUTHORITATIVE, *The Century Magazine* epitomized gentility. It published works of history, fiction, and science that the cultured set deemed compulsory reading, and it portrayed itself as an antidote to the asinine yellow press. The editors cherished accuracy; they deemed even a typo a major blunder and often sent draft articles to experts for fact-checking. In the case of the photos from Alianza, as I was amused to discover when decoding cryptic handwriting in microfilmed correspondence held at Columbia University, the magazine's choice of peer reviewer was someone whose own ideas about Mars were questionable: Nikola Tesla. "The article is very interesting. I was glad to read it," wrote the inventor, who was a close friend of *The Century*'s associate editor. "You may not realize that the photographs are of immense importance as [they are] the first ones settling all doubt about markings," Tesla appended. "As to life [on Mars,] if you will find some millionaire who will *listen* to me we shall know more."

As for the millionaire Wall Street financier J. P. Morgan, he had more important matters than Mars to deal with in October 1907, for he was busy engineering a fix to the nation's banking crisis. Morgan had by now definitively refused to provide additional funds for Tesla's communications tower on Long Island, and Tesla, the despondent genius, could be seen of an evening eating alone, his lanky frame dwarfed by the marble columns and pendulous chandeliers of the Waldorf-Astoria's grand dining hall. Staring into his plate, he gesticulated to no one as he sketched invisible notes on the tablecloth with his cutlery.

WALDORF-ASTORIA HOTEL, NEW YORK.

"Tesla has not appeared in the newspapers for so long a time that the public has forgotten his existence," a commentator remarked in the summer of 1907. Tesla was running out of money and the broader economy was struggling, yet he refused to relinquish his dreams. He reasserted himself on the public stage as he sought new investors. In interviews and articles and letters to the editor, he once again touted his remote-controlled torpedo boat, the prospects of wireless energy, his hope of communicating with Mars. "Signalling to that planet presents itself as a preëminently practical proposition which, to carry out, no human sacrifice could be too great," Tesla opined.

The idea of talking to Martians by wireless seemed less outlandish now than when Tesla first claimed, on the very eve of 1901, that he had received interplanetary messages. In the six years that had passed, wireless technology had advanced considerably and displayed almost limitless potential. Passengers on transatlantic steamers could stay

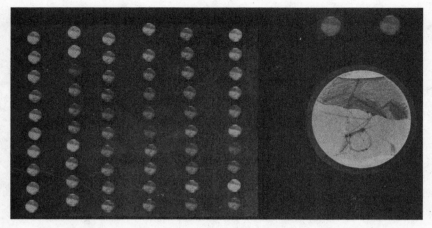

TAKEN JULY 13, WHEN MARS WAS NEAREST THE EARTH.

PROFESSOR LOWELL'S DRAWING OF HIS CORRESPONDING VISUAL OBSERVATION AT THE SAME DATE.

current on politics and the stock market via morning newspapers published onboard, filled with information conveyed through the ether overnight by Marconi. Rival inventors, meanwhile, showed that radio waves could transmit not only the dots and dashes of Morse code but also speech and music.

Now further bolstering Tesla's case for Mars communication were Lowell's new photographs that, like an early holiday gift, hit newsstands in late November, in the Christmas issue of *The Century*. The magazine printed more than three hundred of the tiny, circular pictures, which ran across the pages as if a machine gunner at target practice had shot holes through the paper. "Now, the importance of these little round disks, doubt-killing bullets from the planet of war, is that they reveal to laymen and astronomers alike that markings exist on Mars which cannot be explained on any other supposition than that life able to fashion them is present there at this very moment," Lowell wrote with his usual self-confidence in the accompanying text.

In truth, the printed photographs revealed no such thing. The small, grainy images had been reproduced on the page through the halftone process, which converted any continuous shapes into dots, so it was impossible to conclude—based on what appeared in the magazine—that Mars displayed straight, unbroken lines. To help *The Century*'s readers interpret the tiny pictures, Lowell provided larger drawings of Mars on which he rendered the canals unambiguously, and the magazine placed these beside the photographs for comparison. It was left to the reader to trust that what Lowell sketched and what the camera recorded really were the same, but that seemed a safe assumption given *The Century*'s reputation and Lowell's renown.

...

IN MID-DECEMBER, A BLIZZARD once again struck Boston, the snow riding in on a gale, followed by sleet and rain. The heavy accumulation yanked down telephone wires and crippled transportation while turning the streets to slush, but Lowell, back in Massachusetts, surely felt warmed by the events of the year concluding. The Lowell Expedition to the Andes had distinctly enhanced his standing. The newly famous Mars photographs, both from Chile and Arizona, were now prominently displayed at MIT and would soon travel to New York's hallowed American Museum of Natural History. Meanwhile, the National Geographic Society, which had so publicly fêted Commander Peary after his return from the Arctic in late 1906, honored Lowell with an invitation to lecture about Mars. The request had come from the society's past president, none other than Alexander Graham Bell, who had fallen under the astronomer's spell.

Then, as Lowell's annus mirabilis drew to a close, *The Wall Street Journal* published an editorial on its front page. "Look back upon the year 1907 and pick out what has been, to your mind, the most extraordinary event of the twelve months," it began. "Certainly it has not been the financial panic which is occupying our minds to the exclusion of most other thoughts. That, after all, is a mere temporary disturbance, a mere passing cloud." This hidebound periodical, a newspaper for the sober and phlegmatic, concluded instead that the most extraordinary development of 1907 had been "the proof afforded by the astronomical observations of the year that conscious, intelligent human life exists upon the planet Mars."

Lowell had his staff clip the editorial and paste it in his scrapbook. The people of Mars, and the Martians of Earth, had attained not just credibility but respectability.

...

MORE THAN A CENTURY later, in 2019, I went in search of Alianza.

At 8:00 A.M. on a mild day in late June—winter in the Southern Hemisphere—I departed my hotel in Iquique, now a modern city dotted with high-rises, and steered my compact rental car toward the imposing Atacama. Chile's nitrate industry, once so dominant here, died long ago after chemists in Germany devised a more efficient method for producing fertilizer and explosives, from nitrogen in the atmosphere. Where the railway used to weave its way up the sand cliffs to the mines, a paved road now stretches, and I pressed my Volkswagen up the long slope and onto the high desert plain. Emerging at the top, I was astonished to find the scene precisely as the Todds had described it—unearthly. Indeed, it looked eerily like Mars: gentle hills and escarpments tinged red, all rock and sand, with no signs of life. The morning light emphasized the undulations in the land, painting it with broad patches of shadow.

After an hour's drive inland, I turned right on Route 5, Chile's great north–south ribbon of asphalt—the Panamericana—which runs most of the length of this gangly nation. In this region, the two-lane highway sees only the occasional vehicle as it traverses sheer desolation. The terrain looked like it had been the scene of battle—perhaps trench warfare—cratered and boulder-strewn as if the earth had been turned inside out, which it had, pickaxed and dynamited to get at the precious subterranean layer of nitrates. Continuing south, along an extended stretch of straight road, I at last came upon a rusting sign: OFICINA ALIANZA. I angled off the pavement and onto a dirt path.

To call what I found a ghost town would be generous. It was more akin to an archaeological site. The avenues of barracks that had once housed workers had crumbled to piles of plaster and metal and frayed wooden planks. I pulled over, exited the car, and entered the desert's blazing stillness. I felt as if I had stumbled into a theater long after the drama had concluded and the actors had departed. All was silence and void.

I had brought copies of several photographs I had found among David and Mabel Todd's personal papers, now stored at Yale University. They showed the great Amherst telescope where it had been erected in 1907, beside the mine managers' bungalow. No such dwelling remained, but I looked around for where it once had stood. I noticed a rising slope and, at the top of it, an exterior staircase—five steps that led to a broad, flat area of dirt. Walking over and examining the photos, I concluded that this was the earthen platform upon which the house had rested.

As I paced, my footsteps kicking up dust, I found—along one side of the building's foundation—a low wall. It continued, tracing a perfect rectangle that surrounded a level area: the tennis court. Here was the very spot where Mabel Todd had gazed at Mars and cried. In the distance, I spied a shimmering lake on the horizon. It glittered under a cloudless sky. I knew, however, that this beautiful spectacle was mere mirage. There was no life-giving water, just dazzling illusion.

Part Three

THE EARTHLINGS RESPOND

1908–1916

MARTIANS GAZING AT THE EARTH.

WHAT THE MARTIANS MAY BE DOING AT THIS MOMENT—WITH THE AID OF HUGE TELESCOPIC MECHANISMS—OBSERVING THE LIFE IN THE STREETS OF A EUROPEAN CAPITAL.

CHAPTER
15

Articles of Faith

1908–1909

We all comprehend the world through concentric layers of trust. No single individual can master every domain of knowledge, so we must rely on others to shape our own ideas of what is right, what is factual. Scientific experts, vetted by their peers and published in responsible journals, inform us of the workings of nature. Respected members of our social circles point us toward those ideas and people we should deem most credible. New concepts filter through this mesh of gatekeepers, who help us extract truth from false rumor. Despite these safeguards, however, the system can go awry.

"Men, it has been well said, think in herds; it will be seen that they go mad in herds," the Scottish journalist and author Charles Mackay wrote in 1841 in his classic study of mass hysteria, *Memoirs of Extraordinary Popular Delusions*. Mackay's book is best remembered for its amusing anecdotes of quacks and manipulators, charlatans and impostors, and the great crowds of dupes who have been swept up in epidemics of folly—such as seventeenth-century Holland's tulip mania, during which a single rare bulb reportedly sold for the sum of "4600 florins, a new carriage, two grey horses, and a complete suit of harness" before the speculative bubble burst. But mass delusions require no hoax to begin, no spur of greed to accelerate. Mackay also wrote of the centuries-long history of alchemy, pursued by respected and well-intentioned men who sought to concoct an elixir of life, a potion to confer eternal youth. The public embraced the enterprise because it promised salvation from suffering, not because of a bad actor's purposeful deception. In other words, people believed in alchemy because they *craved* to believe. That same impulse became evident in the growing embrace of life on Mars.

Percival Lowell was surely no hoaxer. What he had seen with his eyes and perceived with his mind he fully accepted as true. Now, having presented the latest Mars photographs as definitive proof of a Martian civilization—and with the imprimatur, no less, of *The Century Magazine*, *The New York Times*, and *The Wall Street Journal*—he accumulated a passel of notable supporters. By 1908, his American adherents included not only Alexander Graham Bell but also Ivy League academics: a Brown University sociologist ("On Mars we can . . . see with our own eyes a race of vast antiquity and supreme wisdom") and a Yale paleontologist ("Since reading your book on the canals I have scarcely been able to think of anything else"). Among Lowell's influential supporters in Britain was Lady Henry Somerset, a women's rights and temperance activist who applauded the evidence ("almost beyond proof") of artificial canals on Mars.

Even those who did *not* believe in Lowell's theory acknowledged that they were losing the battle of public opinion. "To us, it is a joke, but to a great portion of the people, it is a serious matter," Harvard astronomer Edward C. Pickering—an anti-canalist—complained to his colleagues when there was talk, again, of talking to Mars. "If we could get the views of the judges of the Supreme Court of the U.S. so far as they have heard about it, I think they would look upon it as a serious question." Lowell's ideas were spreading well beyond their original domain of astronomy to infect other realms of thought and belief.

. . .

THOMAS PAINE, THE PATRIOT and pamphleteer whose *Common Sense* provided much of the intellectual spark for the American Revolution, abhorred organized religion. In 1793, after the United States had won independence from Great Britain, Paine turned his ire from the king to the clergy. "All national institutions of churches, whether Jewish, Christian, or [Muslim], appear to me no other than human inventions, set up to terrify and enslave mankind," he bravely wrote in *The Age of Reason*, a screed that flung arrows at the teachings of the Bible.

One of Paine's pointed attacks emerged from his assumption that there must be uncountable inhabited planets in the vast universe. Is it not a "strange conceit," he asked with his quill dipped in blasphemy, "that the Almighty, who had millions of worlds equally dependent on his protection, should quit the care of all the rest, and come to die in our world, because, they say, one man and one woman had eaten an apple?" To Paine, the existence of alien life, if proven, would reveal the folly of Christian faith.

A century later, Paine's disciples—America's atheists and freethinkers—resurrected his attack on religion in light of Lowell's evidence that the Martian civilization existed. "Let us suppose that communication be established with Mars and unquestioned proof secured of its habitation, what is to become of the Christian scheme of salvation, and its atonement of blood? It will prove a serious blow to bible cosmogony," declaimed a prominent freethought periodical. "Put the people of earth in communication with the people of Mars, and there will be considerable wriggling in the Christian pulpits."

Contrary to expectations, the faithful did not cower in the face of Lowell's Martians but instead incorporated the extraterrestrials into their preconceived views of the universe. In Ireland, a friar of the Augustinian Order argued that the existence of life elsewhere served to enhance—not diminish—God's majesty, "by enlarging . . . the numbers of those capable of giving glory to the Creator." In New York, a Catholic priest assured his congregants, "The gracious God that guides the feet of man on earth guides too the footsteps of the Martians." An American religious paper, *The Christian Work and the Evangelist*, expressed enthusiasm at the prospect of talking to another world: "We imagine one of the first questions our theologians would ask would be about the first Martian man and woman, and whether they are reported to have behaved as wickedly as ours did."

One Sunday in Kansas City, at All Souls' Unitarian Church, the Reverend Charles Ferguson preached rhapsodically about Lowell and his proof of Martian life:

> What on earth can rival the importance of this news? Consider the implications of it, and how they bear upon our world problems. We see off there in space, marching by our side, a neighbor world, which shows at a glance by impressive signs that it is morally unified and civilized beyond the dreams of our reformers! It is evident that the Martian land is naturally as arid as Sahara and that the desert refuses to yield the necessaries of life save under compulsion of a social intellect that dominates all private interests on the planet, from pole to pole! How would the jangling, class riven nations of the earth fare under such natural conditions?

The congregants, in their Sunday best, likely replied with a hearty *Amen*, for the embrace of Lowell's theories had grown to the point where a commentator remarked in early 1908—nonchalantly, as if stating an obvious fact about public opinion—"We most of us believe . . . that Mars, among other planets, is inhabited."

. . .

LOWELL AND HIS IMAGINED beings seemed to be everywhere by 1908. *The Century*, after printing his Mars photographs in its final issue of 1907, now serialized his most recent Boston lectures, which were soon released in book form as well. Lowell's grandiose comments on the planet's inhabitants—"Their presence certainly ousts us from any unique or self-centered position in the solar system"—infiltrated the news columns of America's dailies just as an alien invaded the funny pages. Mr. Skygack, from Mars—a spindly figure with a pill-shaped head topped by peach fuzz—appeared in a recurring comic drawn by cartoonist A. D. Condo, also known for his popular Diana Dillpickles. Much like Lowell in his younger days, Skygack served as a roving anthropologist, dispatched to Earth to study the ways of humans—

although his bulletins, sent home by wireless, invariably misinterpreted what he saw. (Skygack's puzzled report on a bride and groom at the altar: PAIR WAS PROBABLY GUILTY OF SOME SERIOUS CRIME JUDGING FROM EMOTIONS DEPICTED ON FACES.) The Martian as a naïve observer of human affairs became a trope, a figure of speech. In February 1908, an influential American socialist pleaded for harmony in the fractured labor movement. "I shall consider the subject of Unity the way a traveler from Mars would do—objectively," he said as he launched into a major address framed throughout with this extraterrestrial rhetorical device.

Meanwhile, *Cosmopolitan Magazine*, the same periodical that in 1897 had gripped America with its serialization of *The War of the Worlds*, now entranced readers with an ostensibly factual article called "The Things that Live on Mars." The author, again, was H. G. Wells. "What sort of inhabitants may Mars possess? To this question I gave a certain amount of attention some years ago," began the British writer, whose focus had turned back to the red planet after his 1906 meeting with "my friend, Mr. Percival Lowell." Wells now rejected his earlier conception of Martians as gelatinous, blood-thirsty monsters on stilts. Instead, he offered a new mental picture: "They will probably have heads and eyes and backboned bodies, and ... big shapely skulls." He did not believe, however, that the inhabitants of Mars were human. "As likely as not," he ventured, "they will be covered with feathers or fur" to insulate against the planet's wild temperature swings. Accompanying Wells's words were fantastical illustrations that depicted the Martians much as Lowell presented them—as peaceful, social beings—in this case clustered in cities and sheltering their young. The creatures possessed not only wide eyes and antennae; they sported wings.

Wells's original Martians had looked like demons. Here they were angels.

...

"THERE ARE CERTAIN FEATURES IN WHICH THEY ARE LIKELY TO RESEMBLE US. AND AS LIKELY AS NOT THEY WILL BE COVERED WITH FEATHERS OR FUR. IT IS NO LESS REASONABLE TO SUPPOSE, INSTEAD OF A HAND, A GROUP OF TENTACLES OR PROBOSCIS-LIKE ORGANS."

BELIEF IN THE PEOPLE of Mars was taking on aspects of a pseudo-religion, and if Lowell loomed as its prophet, an apostle soon helped spread the faith through a modern means for disseminating ideas: a national network of women's clubs. From urban centers to farm villages, America's housewives and its female professionals and laborers joined together to exert their power by the hundreds of thousands. Though unable to vote, they used their collective voice to push for better schools, crusade against child labor, establish libraries, and beautify their communities. ("The woman's club movement is one of the most important sociological phenomena of the century—indeed, of all centuries," an early feminist and social reformer noted.) Beyond civic improvement, women's clubs encouraged *self*-improvement. They invited speakers to educate members on scientific, literary, and cultural topics. A favorite lecturer now found herself in especially high demand. "Mabel Loomis Todd . . . is being idolized by the women's clubs," a Boston newspaper noted. "They all wish to hear [her] tell of her trip to the Chilian desert."

Even before she had left for Chile, Mabel Todd had planned an ambitious speaking tour when she returned, and she offered a menu of subjects she might discuss. "We want The Latest News from Mars, surely," read a typical request, from a Massachusetts women's club. Todd's time on the road proved a blur of faces and landscapes and tight connections as she traveled by rail from event to event: Cambridge, Pittsburgh, Boston, Providence, Brooklyn, Milwaukee, Grand Rapids, Philadelphia, Toledo, Buffalo, Chicago, Cleveland, and elsewhere. Through blizzards and illness, she pressed on, delivering more than 150 lectures by the end of 1909.

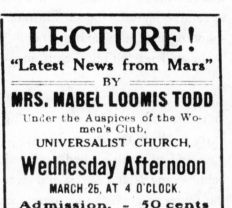

If Percival Lowell appealed to his audiences' intellect and imagination, Mabel Todd aimed for the heart. She won admirers for her elegant outfits and held listeners with her rapid and dramatic delivery.

"That first glimpse of Mars! I sat back with tears rolling down my cheeks, completely overcome with the thought that I had actually seen just what Schiaparelli had been stoutly maintaining against all scoffers since 1877—that is the canals, and many with double lines," she told a women's club in New Jersey, where five hundred people showed up despite a drenching rainstorm.

"We went down to South America absolutely unbiased in our opinions and I came back with the idea that Mars is inhabited by a race of people who are different from us as far as their anatomy is concerned, but who are intelligent and capable of action," she informed another large crowd, this time at a women's church group in Providence, Rhode Island.

Eager to entertain, Mabel Todd rated her performances in her diary. "Talked on Mars till their hair stood on end." "Took them off their feet. It was one of my best." "Room packed almost to breathlessness, and I gave them Mars with all its accompaniments. Tears, embraces, adulation." The reviews from her audiences were similarly effusive. Todd did not force her conclusions about Martian life on her listeners but asked them to decide for themselves if the evidence was convincing. "So completely had she worked out her subject matter, and so logically did she illustrate her points," reported one sober-sounding attendee who heard her in Portland, Maine, "that every one in the audience felt impelled to answer 'yes.'"

...

IT IS NO SURPRISE that the seeming reality of Martian life appealed to the public in this era of existential disorientation. In ages past, religion and superstition had surrounded humankind with enchanted beings—angels and fairies and elves—and had placed the Earth at the center of a small and comprehensible universe, but science had banished those magical creatures and revealed that we float through a cosmos so vast it can terrify. (The French physicist Blaise Pascal famously expressed this deep-seated dread: "The eternal silence of these infinite spaces terrifies me.") Pain and death were the inevitable price of admission to life on Earth, a reality brought home to Garrett P. Serviss when his namesake son, the Olympic athlete, died at twenty-six of a cause unspecified by the newspapers, the young man following his own mother to an untimely grave. Death and ruin preoccupied Camille Flammarion, too. "Disastrous earthquakes . . . have succeeded each other with such extraordinary frequency," he wrote when seismic cataclysms killed thousands in Valparaíso, Chile, and

Kingston, Jamaica, not long after the great San Francisco earthquake of 1906. "We have suffered blow upon blow." The idea that Martians as superior, moral beings might be watching over Earth like guardian angels made people feel less vulnerable, less alone.

A journalist asked David Todd, "Do you think there will ever be communication between the two planets?"

"That is a possibility," he replied. "Several years ago a Frenchman"—and by this he really meant a Frenchwoman, Mme. Guzman—"left a large sum of money to be used to open communication between the Earth and Mars, but I do not know that anyone except Nikola Tesla has ever seriously undertaken the task." In 1909, Earth and Mars would once again head toward a rendezvous, Todd explained: "Mars will be ten percent nearer the Earth next September than in 1907 when I photographed the canals from the Andes Mountains." Indeed, the coming astronomical opposition would bring the planets closer than at any time since Lowell had founded his observatory a decade and a half earlier. Here was a prime opportunity not only to study Mars but, perhaps, to talk to it.

A syndicated article published in newspapers from Syracuse to Pittsburgh to Omaha anticipated the day when the two planets might actually converse, and it helpfully offered, as the headline announced: SOME QUESTIONS MARS MIGHT ANSWER. The list of suggested queries did not address practical matters—how to build an airship, for instance, or dig a canal. Instead, the questions were of a more profound nature: *Where do our spirits go when we die? What is human consciousness? What is life? What is the right way to live?*

"I expect to find an answer to all these world-old questions in the messages from the Martians," David Todd maintained. "Upon Mars they have progressed thousands of years beyond us. All the things that are mysteries here are an open book there." The science of astronomy, which in recent centuries had undermined traditional notions of religion, now seemed poised to restore some of what it had destroyed by cracking the most existential riddles. Lowell and his acolytes had created Martians that were more than angels; they had become stand-ins for God.

CHAPTER
16

Skyward

1908–1909

During 1908, the marvels of the twentieth century grew even more marvelous. Spring saw the opening of the world's newest, tallest skyscraper—New York's Singer Tower, which stretched to over twice the height of the dizzying Flatiron Building. By autumn, the Ford Motor Company introduced the Model T, a budget automobile so affordable, and soon so ubiquitous, that it would transform the nation's very culture and landscape. By December, the American stock market had almost fully rebounded from the panic of the previous year, prompting J. P. Morgan, the bullish financier, to assert, "Any man who is a bear on the future of this country will go broke." But it was Thomas Edison who perhaps best summed up the mood of the era. When *The New York Times* asked him what wonders science and technology might soon achieve, the mythic inventor answered flatly, "Everything, anything, is possible."

The greatest and newest marvel was motorized flight, which had still seemed little more than fantasy when 1908 began. Aviators in the United States and Europe had in recent years constructed experimental airplanes that could lift a man off the ground, but only briefly, and the machines generally proved difficult to maneuver. Two bicycle mechanics from Ohio, Orville and Wilbur Wright, asserted that they had found the secret to sustained, controlled aviation, yet few believed the brothers because even fewer had witnessed their test flights. In August

1908, all skepticism evaporated when Wilbur Wright flew above slack-jawed spectators at a racetrack in the French countryside southwest of Paris, near the ancient city of Le Mans. For almost two minutes, he soared and banked and balletically circled before gently touching down. Two days later, he carved an astonishing figure eight.

WRIGHT CONQUERS AIR, the *New-York Tribune* trumpeted as the miraculous reports crossed the globe, reaching his family back in Dayton. "They treat you in France as if you were a resurrected Columbus," his father gushed in a letter to his son, "and the people gaze as if you had fallen down from Jupiter." Human flight on winged machines became the talk of the age. "Ever since the Wrights came back from Europe everybody has been up in the air," a humorist for the Hearst papers joked. "Balloons have gone out of fashion altogether, and are now only used by scientists who think they can ascend high enough to whistle to Mars."

WILBUR WRIGHT'S AEROPLANE IN FLIGHT NEAR LE MANS.

The newspaper's jest aside, balloons were, in fact, very much in fashion. The lighter-than-air craft, held aloft by hydrogen or coal gas, were vehicles of elite competition, raced by aeronauts who vied for trophies, medals, and hefty purses. Wealthy enthusiasts across Europe and the United States formed aeronautical clubs to promote the sport. Charles S. Rolls, a swashbuckling aristocrat and automotive executive of the Rolls-Royce motorcar company, had recently co-founded the Aero Club of Great Britain. Across the Atlantic, the president of the Aero Club of New England was one Abbott Lawrence Rotch, a prominent Bostonian and Harvard meteorology professor who used weather balloons to probe the atmosphere. In September 1908, these two men met in England to make an ascent over London, and joining them was Rotch's cousin from America, Percival Lowell. Lowell had arrived with a companion of his own. She was a longtime neighbor, a real estate professional who renovated homes, including Lowell's, and had earned the moniker of Boston's "Woman Contractor." Until recently she had been known as Miss Constance Savage Keith. She now went by a new name: Mrs. Percival Lowell.

Earth's lonely Martian was finally partnered. Lowell's decision to wed, at fifty-three, had been typically impulsive, his sudden marriage—and his choice of bride, for she was not of his social class—surprising friends and family. "Prof. Percival Lowell has married a Boston girl," *The Chicago Daily Tribune* japed, "and is reported to be considering the feasibility of taking a honeymoon trip on the canals of Mars." The actual honeymoon—a working vacation, as Lowell's always were—was taken in Europe, where he visited French observatories and hiked a Swiss glacier "to learn more of Earth and more of Mars in the process," he explained. Now in London, Lowell arranged to do what Flammarion had done after his own wedding and carry his new wife skyward.

From their Piccadilly hotel near Buckingham Palace (where King Edward VII was approaching the end of his brief reign), the Lowells, mister and missus, headed out to meet the aeronauts, Messrs. Rolls and

MR. AND MRS. PERCIVAL LOWELL.

Rotch. It was a Saturday on the cusp of autumn but felt like midsummer, and the bulk of London seemed to be enjoying the fine weather, with those of means and leisure preparing to motor in the country or ride the train to the seaside. Through the labyrinth of streets choked with omnibuses and hansom cabs, the Lowells made their way to the north bank of the Thames, in Chelsea. A crowd had assembled around a large balloon that strained upward. When the dapper American groom ushered his bride into the large basket, the spectators, comprising scientists and members of London society, caught their breath. Then, in but a moment, the craft launched and the ground fell away.

No record survives of Lowell's reactions upon flying over London's rooftops much like Mary Poppins, that Edwardian nanny with a magical umbrella, but it must have been enchanting. He quickly climbed above the Earth's most populous city, which covered so large an expanse that it was once suggested the metropolis itself might signal Mars simply by switching its streetlights off and on in unison. From up here, the sailing barges that plied the Thames appeared as toy boats. A mere outstretched palm could hide the great Gothic palace of the Houses of Parliament and the towering dome of the Baroque masterpiece that was St. Paul's Cathedral. With the sweep of an eye, one could take in the entirety of a region comprising 6 million human souls, out to the Surrey Hills, near Woking, where the first of the Martian cylinders crash-landed in *The War of the Worlds*. Even at altitude, the air of London veritably hummed with energy. For Lowell, the symbolism would have been hard to miss. Here he was, the high-flying personage he had been destined to become, a man with the world at his feet.

Lowell's flight was symbolic in another way, though not in a manner that he would have appreciated. As Lowell's career soared to new heights, his ego—not inconsiderable even at sea level—was itself distending, like a gas balloon at altitude. Lowell increasingly saw himself as not just an astronomer but a recondite philosopher whose cosmic theories about the birth and death of planets, and of life itself, made him a pioneering evolutionist whose name would be engraved in the pantheon of science. He soon penned an essay in honor of Charles Darwin, and his words—"The chief function of genius is to change the world's point of view"—he clearly intended in praise of himself. Lowell's bloated ego expanded with each piece of gushing fan mail, with every published tribute.

Sailing over London, Lowell drifted north from Chelsea toward the posh neighborhoods of Belgravia and Mayfair. The scene below presented a tangle of narrow lanes and broad avenues. Had Lowell observed the earthly tableau with an open mind, he might have noted that this city, which touted itself as the pinnacle of civilization, had not been laid out in simple, straight lines. Rather, the intelligent beings had built roads to follow the topography and the whims of urban design. There were curves and circles and bends in the avenues, including a prominent jog in Brompton Road on its way to the great shopping emporium that was Harrods. But Lowell seems to have been uninterested in the confusing maze of streets below him. This we know because he instead focused his attention on a polygon of greenery just ahead—Hyde Park—where a network of linear footpaths met and diverged with precision. To him, this was what intelligent design looked like, an analogue of the terrain he perceived on Mars. He photographed the view and, as soon as he returned to the United States, published the images, rushing them into print in his latest book as further evidence to support his hardened theory.

HYDE PARK AND THE SERPENTINE SHOWING ARTIFICIAL MARKINGS OF EARTH SEEN FROM SPACE. (PHOTOGRAPHED AT 2,200 FEET BY PROFESSORS ROTCH AND LOWELL.)

...

DURING THE MONTH IN which Lowell floated above London, the Wright brothers had continued to astonish the public, demolishing aviation records in quick succession. While Wilbur Wright remained in France, Orville Wright was at Fort Myer, Virginia, which overlooked Washington, D.C., beside Arlington National Cemetery. It was there that he demonstrated his biplane's capabilities to the U.S. Army. On the morning of September 9, he flew for a stunning fifty-seven minutes—by far the longest flight achieved to that date—and then, that very evening, sailed for yet five minutes more. Over the next week, he set three more world records before misfortune struck, when a propeller blade snapped midair, sending his machine in a dive that left him bloodied and fractured in the twisted wreckage while killing his passenger, U.S. Army Lieutenant Thomas Selfridge. Even tragedy could not stall the Wright brothers' momentum, however. On September 21, in France, Wilbur Wright shattered the endurance record again when he flew for a full hour and a half. Three months later, on the very last day of 1908, he remained airborne for some two hours, twenty minutes—an incredible feat. An American journalist declared, "We must surely account the demonstration of the existence of inhabitants in Mars by Prof. Percival Lowell and the ocular proof by the Wright Brothers of the possibility of the continued flight for hours of machines heavier than air to be the two most brilliant achievements of the twentieth century."

The age of aviation had arrived on Earth, and what happened on Earth seemed always to be projected onto Mars. In previous years, the public had at least jokingly envisaged the Martians pedaling bicycles and traveling by motorcar; now the aliens were piloting aircraft. In a feature for the Sunday *New York Herald*, the astronomy writer Mary Proctor proposed that the advanced alien race must "long since have mastered the intricacies" of mechanical flight. "Wealthy

Martians probably own their own airships," she fancifully surmised, "navigating the air at will when calling on their friends."

...

AT THIS TIME WHEN impossibilities were becoming realities—buildings touching the clouds, humans soaring on fabric wings—frenzied speculation of a familiar kind returned. "The demand of the hour seems to be to get into communication with Mars," Lilian Whiting opened her weekly newspaper column in mid-May 1909, when the red planet was nearing its close approach to Earth. "As to the method, one savant differeth from another."

Scientists now proposed a slew of signaling ideas. An astronomer in upstate New York suggested that earthlings flash messages at Mars using a great expanse of electric lights. "Let these signals be repeated every night for several weeks before, during, and after opposition," he wrote. A professor at Johns Hopkins offered the inverse scheme—using *dark* signals on a *light* background in *daytime*. "A large black spot upon the white alkali plains [of a desert]," he said, could be winked at Mars by using vast "sections of black cloth arranged to roll up on long cylinders, exposing the white ground underneath." Yet another method came from Harvard's William H. Pickering, the astronomer who long ago helped Lowell found his observatory in Flagstaff. "The signaling by means of mirrors which I have outlined is practical—the only practical means, I think," he told *The New York Times*.

Pickering's plan called for an enormous field of mirrors, covering a quarter mile square, that could direct the sun's rays heavenward. "Looking down from Mars this reflection would appear like a small point of light upon the surface of the earth," he told the Boston *Sunday Post*. He explained that this mechanism could send flashes at irregular intervals, like the dots and dashes of Morse code. "I have no doubt that providing there were intelligent people on Mars the light would at once attract much attention and would lead eventually to an answering signal." Pickering estimated construction at $10 million, a sum so exorbitant that even he predicted no one would finance it, although for a brief moment it appeared as if some wealthy

Texans just might take up the cause.

These communications schemes were, for the most part, mere thought experiments. Little effort was made to pursue them, and their plausibility was easily questioned. During the upcoming planetary approach, the sun, Earth, and Mars would align, which meant that from the perspective of a Martian looking earthward, our planet would appear as a tiny black dot set against the sun's blinding glare. Visual signals, such as flashing lights, would be nearly impossible to see under such conditions. What is more, the Earth at that time would present its *nightside* to Mars, so any daytime signals using solar mirrors or black cloth in a white desert would be hidden from view until the Earth moved along in its orbit, farther from Mars. A more realistic—or less unrealistic—proposal for signaling was by means of wireless, the method long suggested by Nikola Tesla and now embraced by the always resourceful David Todd.

PROFESSOR PICKERING'S PLAN FOR COMMUNICATING WITH MARS.

...

TODD, THE AMHERST COLLEGE astronomer, had long been interested in human flight and was a member of the Aero Club of New England. He often attended aviation-related social functions, including a celebratory luncheon with the Wright brothers when they visited Washington, D.C., to receive gold medals from America's new president, William Howard Taft. At a recent aeronautical banquet, Todd mentioned that he wished he could soar high in the atmosphere to intercept any radio signals that might come from Mars, ones so faint that they could not penetrate to the Earth's surface. Leo Stevens, one of America's foremost balloon builders, chimed in: *If ten miles would cut much ice, I can rig up an aerial outfit that would carry us there.* The men talked, shook hands, and began planning a dangerous mission to lift a radio receiver to an unprecedented altitude.

They soon disclosed their intentions to the press. "Ten miles is higher than man has ever been," Stevens told the *New-York Tribune* in what was a marked understatement. One of the two highest balloon ascensions to date had carried a pair of British adventurers, in 1862, to an altitude of six (perhaps seven) miles, at which point oxygen deprivation blinded, paralyzed, and nearly killed them. Four decades later, in 1901, two German meteorologists—Arthur Berson and Reinhard Süring—climbed to about the same altitude (six and a half miles), where, despite having brought along tanked oxygen, the men grew listless and drowsy, their breathing tubes fell from their mouths, and they, too, lapsed into unconsciousness in the minus-forty-degree air before a rapid descent saved them.

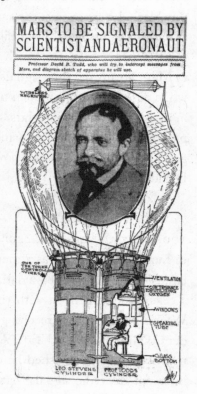

With Leo Stevens and David Todd now hoping to fly several miles higher than either the Britons or Germans, Stevens gamely predicted that "we shall come down alive," for his plan was to build a safer craft, akin to a space capsule, with each man enclosed in his own pressurized aluminum cylinder. "We will have little observation windows and I guess that when we get nine miles in the air the earth will look like a ball to us," Stevens explained.

Todd added wryly, "If either of the tanks should give way at a high altitude the man inside would be a gone goose."

For warmth, the men planned to clothe themselves as if they were climbing the Alps: silk underwear, wool shirts and pants, rubber coats, and caps that pulled down over the ears and face with openings only for eyes, nose, and mouth. They would pack hot and high-calorie foods: chicken sandwiches, hardboiled eggs, thermoses of steaming coffee. And they would prepare for the mission with a training regimen: breathing tanked oxygen on the ground and taking test flights at low altitude. Stevens predicted confidently, "We will be talking to the people of Mars before the 15th of next September."

Some of the statements reported in the press had clearly been embellished by Stevens and Todd to garner attention, and others were possibly embroidered further by the journalists themselves. When Percival Lowell's cousin Lawrence Rotch, the aeronaut and meteorologist, learned about the plans in the New York *Sun*, he wrote to David Todd with a bit of caution, advising that it would be nearly impossible for a hydrogen balloon to carry the two men and their equipment so high. "Berson and Süring, in that record ascension, had a balloon of 300,000 cubic feet," which was far larger than any American balloon, Rotch explained.

Todd, in reply, blamed the press for exaggerating. "Of course you don't believe all that has been spread broad cast about this business," he wrote. "Neither Stevens nor I am at all responsible. If I get up even five miles, I shall get more than all the data I originally desired." Nonetheless, a five-mile ascent would be dangerous and difficult enough.

Headlines from New York to New Zealand touted the audacious stunt, which one commentator dismissed as "mere balloonacy" but others praised as an act of valor in the name of science. "There is . . . every reason to believe that the Martians long ago discovered means of sending us messages which we, however, have been unable to perceive," Camille Flammarion said when asked for comment. "Professor Todd's proposal to ascend as high as possible in order to escape the disturbances of the atmosphere for the purpose of experiment is natural and logical."

It was then announced that David Todd's wife would join her husband on his training flights.

"I suppose you and the Professor will be off to Mars, soon, and I hasten to try to catch you before your aero-car is off in space," a delighted Lilian Whiting wrote Mabel Loomis Todd. "What wonderful times we are living in! Every day is a new revelation."

...

BY THE SUMMER OF 1909, the Wright brothers were shifting their efforts earthward—less time in the air to spend more in court, defending their patents—but the Todds, at last, were ready to proceed aloft. On a Wednesday toward the middle of August, as afternoon shadows lengthened in the countryside of North Central Massachusetts, a great pancake of varnished cloth and netting lay piled in a field. Workers connected it to a gas pipe, and the formless lump began to stir, then bulge. It soon grew into a balloon emblazoned with its name: the *Boston*, owned by the Aero Club of New England. Though no high-altitude craft outfitted with aluminum capsules, it would carry David and Mabel Todd on their inaugural flight, a mile above sea level, to help them develop their air legs.

Eager onlookers arrived by carriage and automobile, and soon a crowd of five thousand had assembled. Mabel and David Todd posed for pictures, she in a green traveling suit and stylish bonnet, he in a jacket and tie. The *Boston* pulled upward and swayed in the wind while two dozen men gripped its long drag rope. When ready for launch, the pilot instructed Mabel Todd, "Get in, now," which she did to great cheers and applause, throwing herself

"JUST OFF!" DAVID AND MABEL TODD'S INAUGURAL FLIGHT.

over the rim and into the woven basket. David Todd climbed in to join her. "Let go!" the pilot shouted. The basket and its occupants shot up, a blur rising into the evening sky.

Almost two years had now elapsed since the Todds returned from Chile with the photographs that Lowell had portrayed as the ultimate proof of intelligent Martian life, thereby shifting public opinion his way. To the average American, Lowell was an expert and a celebrity, arguably the most famous astronomer in the country, a man to be heeded. To other astronomers, however, he remained a gadfly, and a whispering campaign began to build among his critics.

"When you visit me, I want to talk with you about the Lowell situation," the Lick Observatory's then director, William Wallace Campbell, wrote to his counterpart at the new Mount Wilson Observatory, outside Los Angeles. "You have of course noticed that Lowell, the past year or two, has been making much ado in public, and in many matters quite unprofessionally." Campbell asserted that the photographs by Lowell and Todd revealed only the biggest, most obvious of Martian features—no network of fine canals as Lowell portrayed them. ("I fully share your opinion," came the reply from Mount Wilson.) "I think Lowell and Todd are going to be a trial to sane astronomers shortly," Campbell concluded his letter. "My question is just how far they should be allowed to go before somebody steps on their rope."

CHAPTER
17

Endgame

August–December 1909

By late August 1909, Mars commanded attention as it rose each evening in the eastern sky. Night after night, it blazed ever more conspicuously, like a hot coal in Pisces, brightening as it fell into alignment opposite the sun. The precise moment of astronomical opposition—the time of close planetary approach, when one might expect the Martians to toss a neighborly greeting earthward—would occur on September 24, but a few weeks beforehand terrestrial news stole the public's focus. Commander Robert Peary, the relentless and toeless Arctic explorer, was heading south from his latest expedition. "Stars and stripes nailed to North Pole," he telegraphed the Associated Press from Labrador on September 6. *The New York Times* veritably shouted the news with pride: PEARY DISCOVERS THE NORTH POLE AFTER EIGHT TRIALS IN 23 YEARS.

Thrilling as the news would have been under any circumstances, it was doubly astonishing because just five days earlier another American explorer returning from the Arctic had cabled that *he* had reached the Earth's very top, indeed before Peary. This rival was Dr. Frederick A. Cook, a Brooklyn physician and charismatic adventurer with a tongue so silver it was said that if he claimed to have come from Mars, the public would believe him. Cook had gained fame in 1906 when he announced that he had made the first ascent of North America's highest summit—the massive, glaciated Mount McKinley, in Alaska—a feat for which he provided photographic proof. Now in 1909, both Cook and Peary said they had discovered the pole, one of the last firsts of Earth exploration. Given the scuffle over this coveted prize, the press sought the adjudication of other scientists and explorers. "I have always been a staunch admirer of Commander Peary and in my opinion he has done wonderful things among the ice and snow of the Arctic Circle," Percival Lowell said, reflecting a widespread view. "However, as far as we know now, Dr. Cook is the one who will be accorded the crown."

Cook and Peary had been close allies in younger days—Cook had served as surgeon on Peary's expedition to Greenland in 1891—but now they were enemies, and the fight over who would be knighted discoverer of the planet's northernmost point descended into an internecine brawl. Over the coming months, Peary and his backers would launch a campaign to expose Cook as a liar and a cheat. "Do not trouble about Cook's story. . . . He has not been at the pole," Peary cautioned the press. "He has simply handed the public a gold brick"—a swindle, in the slang of the era. This feud dominated headlines and resonated with another intensifying battle.

"Next to the wrangle over the discovery of the Pole in the field of science is the question as to the habitability of the planet Mars," the Rochester *Democrat and Chronicle* commented, noting a backlash against Percival Lowell. "There has been no intimation that a Martian gold brick has been palmed off upon the public," the paper acknowledged. "But the hottest stage of the controversy may not have been reached yet."

...

TO THOSE WHO HAD been paying attention, signs had accumulated that the simmering Mars debate was headed toward an inevitable boil.

The rhetoric flung by Lowell's critics exhibited ever greater vitriol. A noted Philadelphia chemist, in an address to a local scientific society, assailed Lowell's actions as "those of a charlatan" for portraying the thinnest of evidence as if it were proof of a Martian civilization. A geologist at the University of Wisconsin, writing in *Science* magazine, charged Lowell with foisting pseudoscience on a trusting public and declared, "Censure can hardly be too severe upon a man who so unscrupulously deceives." A prominent British astronomer, in a lecture to children, blamed not just Lowell for the Mars frenzy but also the sensation-loving press, a gullible public, and even Mme. Guzman for the generous prize she had endowed to encourage communication with another world.

A turning point in the scientific dispute arrived in August 1909, when America's top stargazers gathered for the annual meeting of the Astronomical and Astrophysical Society of America (now called the American Astronomical Society). The scientists convened in southern Wisconsin, at an imposing building that looked like a cathedral, laid out in the shape of a Latin cross and crowned by a colossal dome. It was instead a shrine to astrophysics: the Yerkes Observatory, which sat along Lake Geneva, a stylish resort for Chicago's elite and a bucolic setting for the intellectual gathering.

Between scientific sessions, attendees enjoyed excursions on the water and luncheons of star-shaped sandwiches at tables decorated with moon vines and sunflowers, yet a certain disquiet ruffled the calm. The astronomical society had been founded a decade earlier to bring order to the still-professionalizing field, and there was now chaos to be addressed. The association's president broached this

YERKES OBSERVATORY.

unpleasant subject on the opening day of the meeting. "The public should not be deceived into believing that we will ever be able to signal to or talk with Mars," he declaimed. "It's the veriest rot."

Lowell's Mars pictures, those "doubt-killing bullets," had not slain all skepticism about the lines on the planet, despite Lowell's attempt to convince the public otherwise. On the contrary, when Lowell sent one of his photographic plates for inspection here to Lake Geneva, "no one at Yerkes could see any lines on it," an astronomer later divulged. "Our conclusion was that Lowell was using his imagination." Meanwhile, a leading American scientist replicated Walter Maunder's "small boy" experiment, and in this case the subjects were not schoolchildren but grown men—in fact, experienced astronomers—who were asked to copy sketches seen at a distance. They, too, were easily fooled into drawing lines where none existed. The idea that Mars was playing tricks on the eye gained new currency.

At Lake Geneva, Lowell's American critics rose up as one. On the last day of the meeting, the society's president, Edward C. Pickering, formally addressed the Mars controversy. For him, it was a subject that hit especially close to home. As head of the Harvard College Observatory, Pickering was the man who had loaned Lowell some of his own staff to establish the observatory in Flagstaff in 1894; it was he who at that time had wished Lowell "all success in an enterprise which promises such valuable and interesting results." Now Pickering regretted what those "interesting results" had spawned. Moreover, it was his own younger brother who had proposed that one might signal Mars using a giant farm of mirrors.

The elder Pickering now spoke to the men and women assembled at Yerkes and argued that the outrageous stories about Mars could undermine the entire field of astronomy, which relied on philanthropy for its funding. "If any of us should go to a wealthy man to ask money for astronomical purposes, we should be very likely to have our wisdom and good judgment doubted by the expression that has appeared in papers that astronomers are attempting to [signal the Martians]." Pickering accused the press of bamboozling the public about the likelihood of talking with Mars. He urged the astronomical society to go on record denouncing such foolish ideas.

The body of astronomers did just that, authorizing a brief but unequivocal statement to the newspapers: "As the public through the misrepresentation of the views of certain astronomers has formed the impression that communication with other planets is at present possible, the Astronomical and Astrophysical Society of America desires to express its belief that all such proposals fall outside the range of sober contemporary science." The language appeared formal and restrained, but it amounted to a harsh rebuke, meant to silence not just journalists but also those scientists who had fueled the fever dreams of the yellow press.

...

NEITHER PERCIVAL LOWELL NOR David Todd attended the meeting at Lake Geneva, but Todd knew beforehand that a tongue-lashing was in store, and he began to distance himself from Lowell even as he continued to plan his high-altitude balloon voyage. Todd now refuted his earlier statements to the press. "My observation of the 'canals' of Mars on the Andes expedition was not wholly convincing," he told the *Boston Evening Transcript*. To *The New York Times*, he confessed even more shockingly, "I have grave doubts if life of any kind exists on Mars."

A reader might well have found Todd's true beliefs about Mars hard to decipher. I, too, was befuddled by Todd's contradictory statements, which he seemed to tailor to specific audiences: at one moment aiming to please his Martian-obsessed benefactor, the next trying to placate his scolding colleagues. Todd coveted public attention and often fed sensational material to the press, but now—chastised—he insisted that the main purpose of his high-altitude balloon flight was not to listen for Martian signals, which he doubted he would receive. Rather, it was to test a breathing apparatus that astronomers might someday use to climb lofty peaks and make telescopic observations in the thin air. He soon abandoned the balloon experiment altogether, blaming a lack of funds.

Where David Todd evinced a flexibility of opinion—or perhaps it was a lack of conviction—Lowell only hardened his stance in the face of pressure from his peers. He answered his critics with new evidence and new arguments. When a team from the Lick Observatory announced that its spectrographic studies detected no evidence of water vapor in the Martian atmosphere, raising questions about the planet's habitability, Lowell replied that a similar investigation by his observatory *had* found water vapor on Mars—*and life-giving oxygen too*. The vaunted British naturalist Alfred Russel Wallace, who had conceived the theory of evolution by natural selection contemporaneously with Charles Darwin, had also become a Lowell assailant. He ridiculed the very idea of irrigation channels on a desert planet because so much water would be lost to the air—"The mere attempt to use open canals for such a purpose shows complete ignorance and stupidity in these alleged very superior beings," Wallace wrote—to which Lowell offered a simple rebuttal: The canals, he now suggested, "would probably be covered to prevent evaporation."

Lowell proved a deft opponent, ready to defend himself against all incoming attacks. He seemed to enjoy the fight as if he were competing in a debate tournament or playing a game of chess. Few combatants could match his wit and skill.

For now, however, the Mars debate had been pushed aside by the more pressing North Pole controversy. On September 19, at newsstands across Paris, the illustrated cover of a Sunday supplement theatrically

LA CONQUÊTE DU PÔLE NORD
Le docteur Cook et le commandant Peary s'en disputent la gloire

imagined the feud between Dr. Frederick Cook and Commander Robert Peary, depicting them literally at fisticuffs at the frozen top of the Earth before an incongruous crowd of penguin onlookers. "To which of the two Americans will belong the glory of having discovered the pole?" the paper asked. The public watched the drama with fascination and revulsion. Exploration was considered a gentleman's sport, the competitors presumed to be upstanding and honest, but here it had devolved into a street fight. Peary insisted that his rival could not have traveled far enough north on the sea ice to reach the pole based on what Cook's Inuit helpers had said of the journey. "The Eskimos laughed at Dr. Cook's story," Peary scoffed. Cook in response pledged to offer proof that would eventually vindicate his claim.

The following day—Monday, September 20—the contest over Earth's neighboring planet headed toward a similar one-on-one duel, when a keen astronomer with pugilistic tendencies passed those Paris newsstands on his way to a rendezvous with Mars.

...

IN THE CASE OF any doctrine, religious or scientific, the greatest threat to its survival is often the disillusioned convert, because it is the former believer who possesses the credibility, insider knowledge, and motivation to bring the edifice down.

Eugène-Michel Antoniadi, one might say, had been baptized in the canalist church, for as a young man, in 1893, he had moved from Turkey to France to apprentice as an astronomer under Camille Flammarion. Gazing at Mars through the telescope at Juvisy, Antoniadi at first struggled to discern the canals, but with experience he soon began drafting elaborate maps of the planet cobwebbed with lines. "Had it not been for Prof. Schiaparelli's wonderful discoveries, and the foreknowledge that 'the canals are there,' [I] would have missed three-quarters at least of those seen now," he wrote in 1898. Still, some doubt lingered, and in 1903—when Walter Maunder published his "small boy theory" that suggested the lines on Mars were illusions—Antoniadi reverted to calling himself "agnostic." He returned to the fold a few years later, after viewing Lowell's 1907 Mars pictures. "Regarding the objectivity of the canals of Mars, there seems no necessity or room for doubt after the truly splendid photographic results," he asserted.

MR. E. M. ANTONIADI.

Like Lowell, Antoniadi was a man of wealth and education for whom astronomy was a passion more than a profession. For a time, he stepped away from the stars to pursue his interest in art, a realm in which he differed markedly from Lowell. Lowell proved a crude draftsman; his drawings of Martian landforms were simplistic, stark, and plain, with hard lines that implied certitude. Antoniadi displayed far more comfort with nuance and shades of gray, both in his drawing and in his thinking, and he sketched like a master; he could render the face of a planet with subtlety and realism. In late 1909, he received an invitation to bring his artist's hand and astronomer's eye to Europe's most powerful telescope, to draw portraits of Mars around the time of its opposition.

On September 20, Antoniadi traveled from his home in Paris to the city's southwestern outskirts. He arrived at the Meudon Observatory, with its great ironwork dome that capped the remnants of what had once been a royal château during the reign of Louis XIV.

The setting itself was truly regal: a terraced park that overlooked the Seine and offered a sweeping view toward the Eiffel Tower. The bright and clear day dissolved into misty night as fog settled in the valley below the observatory. Antoniadi entered the building and approached a steel telescope even larger than Lowell's at Flagstaff. It sat on its own enormous mount, like a pedestal, that raised it several stories off the ground.

Antoniadi walked to a small platform against the wall, which served as a kind of crude elevator. Using a system of pulleys, he slowly hoisted himself upward. The creaking wood and chirring rope echoed in the vast chamber. Once he had raised himself to the level of the eyepiece, Antoniadi aimed the great tube through a narrow slit in the dome toward Mars, which burned like a red beacon in the night. He put his pupil to the glass. The planet glowed bright and round. Magnified, it appeared larger than the full moon with the naked eye.

OBSERVATOIRE DE MEUDON.

He had observed Mars over many years through multiple telescopes, but until that moment Antoniadi had only seen the planet the way Lowell and other astronomers almost always saw it—as a jiggling disk that warped and blurred due to the Earth's turbulent atmosphere. Under normal conditions, observing Mars was like staring at a curious red rock at the bottom of a flowing stream, the rippled water distorting the image. On this night, however, the air above Paris was exceptionally stable. It was as if the running water had stilled to a glassy sheen, allowing him a clear inspection of the object on the streambed.

For two magical hours, Mars held dead steady. "Its appearance was stunning," Antoniadi exclaimed the following day. "The view was a revelation!"

At first, Antoniadi sensed that he must be dreaming, or perhaps he had been miraculously transported to one of Mars's tiny moons and was looking down upon the planet from its own near orbit; the view was that perfect. His heart must have raced as he scanned the terrain that he knew so well but now could see with unparalleled definition. He cast his eye toward the places where Schiaparelli and Lowell—and he himself in earlier years—had marked canals as sharp lines on their maps. He hunted down the Thoth and Amenthes, the intersecting waterways that Lowell surmised were managed cooperatively by adjoining irrigation districts. Antoniadi found the Thoth but now could see it resolved into its component parts. It was not a continuous feature but rather "a succession of a few very faint and diffused knots." Where the Amenthes should have been, he observed nothing that even hinted at a line.

The situation was similar across the planet. A canal called the Borbyses turned out to be "a very irregular and curved knotted band." The Nepenthes was not a feature of its own but merely the edge of the shaded region called Libya. He noted complex patterns mottling the landscape—some areas looked marbled or checkered, others spotted like a leopard—but everything appeared natural, not artificial. "Geometry was conspicuous by its complete absence," he concluded. With his right hand, he drew what he observed, and what he observed was no network of canals, neither single nor double.

"When I saw Mars at Meudon," Antoniadi later wrote, "I sighed at all the Flagstaff spider's webs, and thought I should work day and night to demolish such whimsical provocations of truth." It was as if he had experienced a religious vision. He felt duty-bound to preach what the heavens had revealed to his mortal eyes.

Antoniadi was not just an astronomer and artist; he was also a chess master, one of the top players in France. He understood strategy, could think several moves ahead, and relished a good contest. He soon lifted his pen and began to write with care: "Dear Prof. Lowell."

...

PERCIVAL LOWELL WAS IN Arizona, on Mars Hill, where he savored sherry and cream peppermints at the rustic and well-appointed Baronial Mansion, with its music box in the dining room and insects in the peeling walls. When an envelope arrived from Paris, he opened it to find a note in elegant script and perfect English. It began politely, fawningly. "I am extremely indebted to you . . . above all, for the *marvellous* photographs you are taking of the planet Mars," Antoniadi had written, bestowing deference upon his famous American colleague. "Of course, the Flagstaff skies must give you an ideal definition." Antoniadi then heaped yet more praise. Lowell's book *Mars and Its Canals* had recently been translated into French and was now available in Paris. "I shall procure [it] at once and devour," he promised. To one not a party to this conversation—like me, when I read the words on the yellowing, century-old stationery—it seemed an obvious ploy to disarm Lowell with flattery, a wise approach given the fragile ego of the letter's recipient. It was all prelude, however, to Antoniadi's news of his own recent observations of Mars.

"It might interest you to hear that my Meudon work has surpassed all my expectations," he continued. "I hasten, therefore, to send you 4 of my best drawings." Antoniadi enclosed pencil sketches he had made during his exceptional first visit to the observatory and on subsequent nights. "Here, my experience in drawing proved of immense assistance," he attested. "I sat down and drew correctly, both with regard to form and intensity, all the markings visible." He said little more about the illustrations because, he explained, "they speak by themselves to [one] like yourself, who knows the planet better than anybody else." Antoniadi offered them as pieces of astronomical art, as if they were presents, topping each portrait with a gracious "To Mr. Percival Lowell." But these were not gifts; they were grenades. The sketches conveyed an explosive message: Mars was devoid of straight-line canals.

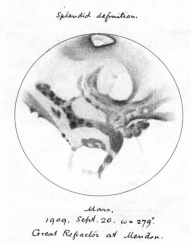

To Mr. Percival Lowell.

Splendid definition.

Mars.
1909. Sept. 20. ω = 279°.
Great Refractor at Meudon.
EM Antoniadi

One can imagine Lowell steeling himself, his cheeks rouging as he stared at the sketches. To him—a man used to subservience—the drawings from the younger, lesser-known European bordered on impudence, but Antoniadi had been outwardly respectful, and Lowell was bound to answer in similar fashion.

"Dear Mr. Antoniadi," Lowell began his typewritten response, "I thank you for the drawings you have kindly sent me." Antoniadi had actually sent a fifth drawing, to compare with the others. It depicted Mars as he saw it under *poor* viewing conditions, when the air above Meudon was agitated. In it, the details of the planet's surface looked smudged, and some of the features fused into thin lines that radiated outward from central points, hinting at the existence of a canal network. Antoniadi had meant for this drawing to show how the so-called canals only *appeared* to exist when a roiling atmosphere caused the underlying details to blur, yet Lowell seized on this sketch. "The one you marked tremulous definition strikes me as the best. It is capital. The others seem not so well defined." Lowell had turned Antoniadi's evidence on its head, focusing on the *least* accurate drawing to support his own theory.

To Mr. Percival Lowell.

Tremulous definition.

Mars.
1909. Sept. 20. ω = 301°.
Great Refractor at Meudon.
EM Antoniadi

Lowell explained, in patronizing fashion, that the Meudon telescope was too powerful to reveal the canals most of the time because the enormous lens accentuated distortion caused by the Earth's atmosphere and garbled the view of the Martian surface. "This is the great danger with a large aperture[:] a seeming superbness of image when in fact there is a fine imperceptible blurring which transforms the detail really continuous into apparent patches." (Lowell had often made this self-serving excuse to explain why astronomers using telescopes larger than his, such as at the Lick and Yerkes Observatories, could not see the canals. The lenses were *too big*, he argued.) To prove his point, Lowell sent Antoniadi some of his own recent sketches, which depicted straight-line canals on the planet. Compared with Antoniadi's masterful drawings, Lowell's looked inartistic, a child's stick-figure Mars.

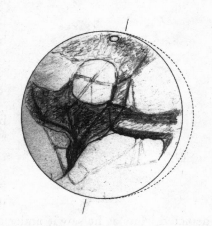

MARS. OCT. 31, 1909.
(P. LOWELL)

One wonders—*I* wondered—whether Antoniadi realistically expected Lowell to respond with an open mind to the drawings from Meudon. Surely not. Antoniadi must have known that Lowell, famously stubborn, would defend his theory against all dissent. Antoniadi likely wrote to Flagstaff not to change Lowell's view but to put him on notice that a challenge was coming. It was a warning shot, and in a subsequent letter from Paris, Antoniadi made his position more explicit.

"Of course, both of us are convinced of having represented, to the best of our ability, what we have seen," he wrote, but "as I have not seen any geometrical canal network, I am inclined to consider it as an optical symbol of a more complex structure of the Martian deserts, whose appearance is quite irregular to my eye. Such a position differs from yours," he added, acknowledging the obvious, then continued: "We both work honestly for the discovery of truth; and we are both eager to accept the truth, even at the hands of a scientific foe, if necessary." It was not at all clear, however, that Lowell was ready to embrace the truth.

Lowell did not immediately respond this time. To him, Antoniadi must have appeared little more than a gnat, a nuisance, but the European amateur had the makings of a dangerous adversary. He was persistent, persuasive, and well connected, much like Lowell himself.

...

ANTONIADI HAD A REASON, and a platform, for circulating his ideas widely. Although he lived in France, he belonged to the British Astronomical Association and served as the director of its Mars Section, so it fell to him to compile and disseminate members' observations of the planet. He shared his revelatory drawings and stark conclusions in a flurry of scientific letters that he mailed across the globe. "These first observations do not confirm the existence of a geometric network of straight lines [on Mars]," he dispassionately informed the French Academy of Sciences. To Britain's Royal Astronomical Society he wrote with more bile: "The spider's webs of Mars are doomed to become a myth of the past." He declared confidently to a prominent American astronomer at the Yerkes Observatory, "The geometrical network of Dr. Lowell is entirely non-existent."

Antoniadi also contacted the distinguished astronomer of Milan whose original observations had inspired Lowell. When writing to Giovanni Schiaparelli about the lack of a canal network on Mars, Antoniadi was careful to express himself diplomatically. "I am convinced that where you have drawn a line there is something on Mars (as far as I can tell)," he explained. "So your 'canals' have an objective basis" even if, as he found, they were neither straight nor *actual* canals. As for the hundreds of additional lines drawn on Mars by Lowell, Antoniadi insisted that they were pure fantasies of the eye and brain.

As the weeks progressed, Antoniadi's attacks grew more public. He debunked Lowell's canal-covered orb in newspapers and magazines in Greece, France, England, and the United States while the battle over the North Pole reached its fiery climax.

· · ·

BY DECEMBER 1909, PUBLIC sentiment had turned against Dr. Cook, who insisted he had reached the North Pole before Commander Peary. First, the guide who had helped Cook's celebrated 1906 ascent of Mount McKinley came forward with a bombshell confession: He and the good doctor, he said, had never reached the mountaintop. Rather, they had staged photographs and faked diary entries to make it *look* like they had reached the summit. As for the claim to the pole, Cook submitted evidence of his feat to an independent panel convened by the University of Copenhagen, which took up the issue because Cook's northward trek had begun from Greenland, an Arctic realm of the Kingdom of Denmark. On December 21, the committee issued its crushing verdict: Cook's evidence was "inexcusably lacking" and offered "no proof" that he had attained the pole.

Individual members of the commission spoke in even harsher terms. They tarred Cook as "an incredibly stupid bungler" if not "a monumental faker [and] deliberate swindler." The once-celebrated explorer, vilified as a fraud, had by now fled for parts unknown to hide in disgrace and disguise. "Maybe the doctor has gone on an exploring expedition to Mars," an Illinois newspaper surmised with a wink. Cook's archrival, Peary, now basked in the accolade he so desperately craved, honored by the National Geographic Society and recognized by the U.S. Congress for having found the pole. (Decades after Peary's death, however, history would turn against him. Experts today judge that Peary, too, was likely a cheat who fell short of the point where longitudes converge.)

One week after the polar commission's ruling from Copenhagen, a meeting in London was set to discuss that other raging planetary controversy, about Mars. On the final Wednesday of 1909, astronomers walked through cold sun to the north bank of the Thames and entered Sion College, a churchlike building with Gothic architecture, stained glass, and carved wood. It was a clubhouse for clergy that offered its meeting hall for monthly gatherings of the British Astronomical Association. This afternoon's assembly would be the valedictory one of the year. Attendees took their seats, and after a few preliminaries—the reading of minutes, the announcement of new candidates for membership—the agenda turned to Mars.

Walter Maunder, the Greenwich astronomer who had presented his "small boy theory" in this very room six years earlier, took the floor. He had been railing against the canal network of both Schiaparelli and Lowell for more than two decades, to little effect. One can imagine the grin traced by his lips as he clutched

SION COLLEGE, NEW BUILDING ON THE THAMES EMBANKMENT.

the report he had received from Paris—from Antoniadi, the head of the association's Mars Section. Antoniadi offered new details of his recent observations and, with them, attempted to bury the Martian canals deep in their grave. "*The true appearance of the planet Mars is a natural one, and comparable to that of the Earth and of the Moon,*" Antoniadi had written. "*Under good seeing there is no trace whatever of a geometrical network.*"

These were mere words, of course, yet there were pictures, too—not just Antoniadi's masterful drawings but also new photographs of Mars taken through the enormous telescope at California's Mount Wilson Observatory. The meeting attendees examined the pictures, which were unquestionably superior to the minuscule, grainy images produced by the Lowell Observatory. These new photos more clearly showed the surface of Mars and revealed no straight-line canals. What they did show were naturalistic streaks and splotches that matched the very shapes Antoniadi had depicted in his sketches, verifying what he claimed to have seen. As for the earlier pictures from Flagstaff and

| E. M. ANTONIADI, | FROM A PHOTOGRAPH, | PROFESSOR LOWELL, |
| MEUDON OBSERVATORY. | MOUNT WILSON. | FLAGSTAFF. |

VIEWS OF MARS (SYRTIS MAJOR AND LACUS MOERIS REGION) IN 1909 WITH VARIOUS TELESCOPES.

Alianza that Lowell had touted so highly, they now seemed no more reliable than Dr. Frederick Cook's fraudulent photographs of his supposed conquest of Mount McKinley.

At the meeting in London, Maunder read aloud Antoniadi's final conclusion about Lowell's canals:

> We shall not expatiate on the circumstances attending the fall of the geometrical network. Yet we cannot refrain from pointing out that it vanished when the planet was practically at its closest approaches to the Earth, high above the horizon, and scrutinized with the best instruments of our time. And the fact that *no straight lines could be held steadily when much more delicate detail was continually visible* constitutes a fatal objection to their crumbling existence. It is thus by a disregard of the dangers of pressing too closely the evidence of our senses that some observers framed those startling theories about the planet. And, whilst Mars became the abode of superhuman channel diggers, reason was gravely insulted.

After reading the report to the assembly, Maunder added a few thoughts of his own about the strange era the Earth had just lived through. "There never was any real ground for supposing that we had in the marking of Mars any evidence of artificial action," he professed. "Had it not been that the idea was a somewhat sensational one, we would never have heard of it, and it is the better for science that it has now been completely disposed of." Maunder then offered a statement that would be carried by newspapers across the globe. "You may sleep quietly in your beds," he said jovially, "without any fear of invasion from Mars."

When a reporter in Boston telephoned Lowell about the astronomical meeting in London and its findings, Lowell replied dismissively: "It doesn't interest me in the least. I am very sorry for them."

To others, however, it seemed a definitive ending.

As 1909 rolled to a close, newspapers reflected on the achievements of the twentieth century to date, now almost a decade in. America and the world had marveled at the advent of mechanical flight, at the discovery of the North Pole, at wireless signals that spanned the oceans. The existence of a Martian civilization, however, had proved too fantastical even for this age of wonders. "Mars has been imposing on the credulity of a trustful world," the *Los Angeles Herald* editorialized on the final day of what had been a remarkable year. "It does not matter who found the pole. Of one scientific fact there can be no question, and that is, there are no canals on Mars."

CHAPTER
18

POETIC ACHIEVEMENT

1910–1916

In truth, the canal controversy had not yet been fully resolved. The Martian civilization that had enthralled the human imagination for almost two decades had been slow to materialize from interplanetary space, and its disappearance would prove equally gradual. But while debates over the markings on Mars would continue for some time, the winds of scientific thought had distinctly shifted. By degrees, Percival Lowell's most ardent supporters fell away, leaving him intellectually isolated.

David Todd had already, in 1909, expressed doubt about the existence of the canals and the aliens that supposedly had built them. Now in the winter of 1910, Garrett P. Serviss told his vast audience of American newsreaders that the latest Mars photographs from the Mount Wilson Observatory constituted "the crowning argument against the 'canals.'" A few months later came a yet more shocking defection.

Giovanni Schiaparelli at this point was in his mid-seventies and too blind to use a telescope, but he actively corresponded with astronomers around the world, and he well understood the resurgent skepticism toward Lowell's theory. Perhaps feeling his own mortality, perhaps attempting to shield his reputation, he wrote to a German science magazine in May of 1910 and conceded that the lines and streaks he drew on Mars may be entirely natural. "[They] teach us nothing at all at present in regard to the probability or improbability of intelligent beings on this planet," he confessed, and he added—further distancing himself both from Lowell and his own younger self—"The name 'canal' should be avoided." The following month, Schiaparelli went to bed and failed to rise. Two weeks later, on a day when the bright sun shone on the cobblestoned streets of the nearby Palazzo di Brera, the ailing geographer of Mars exhaled his last earthly breath.

Lowell authored a heartfelt remembrance of his great mentor, but he refused to follow Schiaparelli into the darkness and acknowledge even the possibly of an uninhabited Mars. On the contrary, Lowell again went on the offensive. He claimed to have discovered two new canals—conspicuous ones, each a thousand miles long, in a place where none had been seen before—and concluded that these waterways must be not only newly *observed* but newly *constructed*, built in a matter of months. Even his supporters found the claim preposterous.

"It is not thinkable that the inhabitants of Mars, however intelligent they may be, could in so short a time dig canals of such magnitude," noted a skeptical Nikola Tesla, who retained his faith in the Martians but was losing trust in Lowell. Flammarion, too, cautioned against Lowell's interpretation of these new lines on Mars: "Let's not rush to attribute them to the engineers of our neighboring humanity."

Lowell fumed at the accumulation of doubters. "The difficulty in establishing the fact that Mars is inhabited lies not in the lack of intelligence on Mars," he railed, "but rather in the lack of it here." Conveying a sense of haughty invincibility, Lowell stubbornly refused to update his thinking when new evidence came in, even as the world of science—and the broader culture—changed almost unrecognizably.

...

AS THE 1910S ADVANCED, the prim manners of the Edwardian era and Victorian age were placed on a pyre and reduced to ashes. In music and books, art and fashion, and in the social mores around sex and gender, iconoclasts set ablaze the genteel traditions of a bygone epoch. A younger generation broke from its starchy past, embracing new forms of self-expression. To literature came the modernist writings of Virginia Woolf and the Bloomsbury Group, to ballet the strange dissonance of Stravinsky's "The Rite of Spring." In painting, the perplexing Cubists and Futurists bewildered the American public at the 1913 Armory Show in New York. "In this recent art exhibition the lunatic fringe was fully in evidence," huffed Theodore Roosevelt, who nonetheless in politics promoted his own new approach, the New Nationalism. Renewal was the watchword.

Old Mars, meanwhile, fell into the distance. By 1912, Mabel Loomis Todd ceased delivering regular lectures about the planet. The following year, Mr. Skygack exited the comics. An editor at a popular magazine offered Percival Lowell some unsolicited advice: "Personally I feel that so much has been written about Mars within the last few years, that it might be better for you to take a new field."

While speculation about life on Mars would continue for decades, and the yellow-press Sunday supplements hardly wearied of the subject, the broader public was moving on, and so was Lowell's circle of associates. One of his oldest friends discreetly confessed to another, *entre nous*, that he had grown tired of the Mars talk. Lowell's siblings, meanwhile, threatened to outshine their luminary brother. Lawrence Lowell, a professor of government, had ascended to one of the highest thrones in academia, the presidency of Harvard. The youngest of the family, Amy Lowell—a Boston eccentric, famous for her love of cigars and her thinly veiled romance with the actress Ada Dwyer Russell—found celebrity as a leader of a new movement in poetry: *free verse*, "what the French call 'Vers Libre,'" Amy Lowell explained.

Science, too, was in the midst of revolutions, which overturned the very principles that underlay Percival Lowell's Martian theory. After Marie and Pierre Curie discovered radium and the enormous energy released by its decay, geologists realized that planets do not indefinitely cool from their initial molten state but generate their own internal heat via radioactivity, an epiphany that caused scientists to revisit their estimates for the age of the Earth and, by extension, Mars. A new breed of anthropologists, meanwhile, was challenging the antiquated (and racist) belief that cultures evolved upward along a hierarchical ladder, from savage to barbarian to civilized. Even the most fundamental assumptions about time and space, and mass and energy, were toppling. "Some of the consequences deducible from the relativity theory are such as may well make the ordinary man stare, either with amazement or with incredulity," Garrett P. Serviss remarked on the radical ideas of Albert Einstein.

America's Martian, however, remained a foppish relic of a dying past. "With fine constancy Percival Lowell continues to spend his fortune on study of the planet Mars," Boston's *Christian Science Monitor* observed. "His brother may administer Harvard and his sister may write vers libre . . . but he steadfastly gathers evidence which convinces him that the Martians are canal building on a grandiose scale."

When Earth swung past the red planet in 1911 and 1914, Lowell continued to see and draw his cobwebs on Mars. Even at his office in the Boston Stock Exchange Building, where he managed his investments, he surrounded himself not only with account ledgers but also with Martian maps and globes, and he talked of the canals to those who would listen. "He leaps from point to point and seems to make no stops at way stations," one visitor remarked on Lowell's rapid-fire delivery. Lowell harangued his guest with declamations. "I have absolute proof that the planet Mars is inhabited," he insisted. "Nothing contradictory to my theory has ever been discovered." His pronouncements now seemed less the visions of a genius and more the rantings of a fool.

In psychiatric terms, Lowell was cycling between mania and depression. "I have been down and up and down," he confessed to his sister Amy, and as his theory fell into disrepute, the downs grew more frequent. "It is nervous exhaustion," his secretary reported. "Some days he cannot even telephone." "The doctor says it is of utmost importance that his *brain* should rest," his wife advised the staff at the observatory, where Lowell's assistants complained of his touchiness and tantrums.

In the spring of 1914, Lowell managed a trip to Europe and joined Camille Flammarion for a solstice celebration atop the Eiffel Tower, but the French astronomer perceived that his American friend was not well. "His health then was not very satisfactory," Flammarion later recalled. "He appeared even to be strangely neurasthenic." Lowell's depression, again blamed on the archaic diagnosis of neurasthenia, had unsurprisingly returned after a decade's absence. "The tendency to relapse increases with each attack," an influential medical text noted of the condition. What is more, according to the era's antiquated science, the symptoms could worsen over time. "Neurasthenia melts by infinite gradations into other morbid groups," the book warned. "It is especially apt to partake of the nature of hysteria [and] delusional insanity."

...

LOWELL REMAINED IN EUROPE through late June and July of 1914, as a geopolitical shock convulsed the continent. The assassination in Sarajevo of Austrian Archduke Franz Ferdinand triggered an inexorable descent into armed conflict, which erupted just as Lowell departed for home. Amy Lowell was still in London when Germany invaded Belgium and France, and she heard the crowds in Piccadilly chant, "We want war! We want war!" As Britain sent its young men to the trenches—where they would be shellshocked, maimed, and dismembered in the cold and blood-soaked muck—she evoked, in prose poetry, the terror of battle. "Boom! The Cathedral is a torch, and the houses next to it begin to scorch. Boom!"

Through 1915 and 1916, as death arrived in new and terrifying ways, Europe devolved into a hellscape. Airplanes and airships rained bombs from the clouds. Flamethrowers incinerated human flesh, and poison gas asphyxiated. Steel-plated tanks, spitting bullets, crushed the injured under rolling treads. These modern means of technological slaughter reminded many of the fictional heat rays, poison smoke, and armored tripods of a novel published almost twenty years earlier. "In those two decades mankind has come up with [H. G.] Wells's Martians," *The New York Times* noted ruefully. Yet in this "present war of the world," as Percival Lowell called it, the existential threat to Earth had emerged not from space but from ourselves.

Amy Lowell lamented:

> Toll the bells in the steeples left standing. Half-mast
> The flags which meant order, for order is past.
> Take the dust of the streets and sprinkle your head,
> The civilization we've worked for is dead.

Distanced from the horrors of World War I, Percival Lowell increasingly hid in Flagstaff like an Old West outlaw. He expanded his observatory—new facilities, more projects—while he built a fortress around his beliefs. "Mars' Hill is a veritable colony, 3 observing domes, 4 astronomers houses, the new building, etc. etc.," he wrote an old college friend who had recently been appointed U.S. ambassador to Argentina. "You, as a diplomat, will understand the style one must keep up as an accredited envoi of Mars to Earth."

Lowell's Mars fixation appeared increasingly delusional, but in this time of unprecedented global warfare and unimaginable despair, sympathetic eyes turned back toward his vision of a world, *any* world, at peace. "Humanity has not, thus far, made such a shining success of its birthright that we need assume that no more rational and purposeful living beings exist than men," *The New York Times* editorialized. "We feel that it would be, on the whole, rather gratifying to be permitted to believe that the wonderfully gifted inhabitants of Professor Lowell's Mars are beings of a better sort." Even if his canal-building aliens were mythical, they remained to many a comforting myth. Lowell still possessed an enduring base of fans, and they welcomed him back to the public stage for a curtain call.

. . .

ON A WEDNESDAY IN late September 1916, while German air raids on Britain made headlines in American dailies (yet still before the United States entered the war), Lowell departed on a lecture tour, his most extensive in years. It would be a three-week trip through the Pacific Northwest and California to address high school auditoriums and college gymnasiums. ("The ear of the next generation is the thing," he had once told his friend Edward S. Morse.) Now sixty-one, Lowell looked even older—the bags beneath his eyes unmasked his years of worry and depression—but as he lectured to students, he seemed rejuvenated. Here were moldable minds, not yet set against him. To them, Lowell was a celebrity, a man worth listening to.

"Our observations have convinced us without a doubt that the lines on Mars which have caused so much discussion, are canals," he told an overflow crowd at Stanford University. "This doesn't mean that the Martians are human beings, like ourselves," he elaborated to thousands of high schoolers in Spokane. "The highest intelligence among the Martians may have developed in any form; it may be that the ants or lizards are the dominant intelligent beings."

In Portland, in a speech titled "Great Discoveries and Their Reception" and with his narcissism on full display, Lowell reflected on the struggles of the lone genius. "The undesired outsider is ignored, poohpoohed, denounced, or all three in one according to circumstances," he cautioned his young listeners, yet he urged them not to let naysayers divert them from the dogged pursuit of scientific knowledge. "Gauge your work by its truth to nature, not by the plaudits it receives from man. In the end, the truth will prevail and though you may never live to see it, your work will be recognized after you are gone."

During his tour, Lowell dined with astronomers from the University of California and his nemesis institution, the Lick Observatory. Although the Lick stargazers still ridiculed Lowell behind his back, he was treated cordially. "Lick and Berkeley have now reached the very respectful stage, shown by their distinguished consideration of the Martian ambassador," he wrote, mistaking politeness for deference. "One could not expect more of mere mortals educated in the Martian dark ages."

Returning to Flagstaff by mid-October, where news of his successful speaking tour topped the local paper, Lowell repaired to Mars Hill to enjoy the changing seasons. Pumpkins from his garden lay heaped in piles. The aspens shone gold on the San Francisco Peaks. A few weeks later, as winter approached, he warmed himself by a fire while his wife, Constance, read aloud from Amy Lowell's latest poetry collection. Basking in the glow of his sister's success and his own return to public life, he seemed content to stay at home for a while. "Universities from Texas to Maine now want lectures but enough is as good as a feast," he wrote his secretary. "Sleet and snow yesterday; blue sky today." The next morning, a blood vessel ruptured in Percival Lowell's brain. He fell unconscious, and before midnight the animating force of his imaginative mind departed his body forever.

Although her husband had not been devout, Constance Lowell arranged for the funeral at Flagstaff's Episcopal church, followed by a graveside service on Mars Hill, where a pastor offered tribute to the fallen astronomer. "Kind friends," the reverend began. "In the language of ancient Rome, which he knew so well and delighted in quoting, we may truly say of our friend departed, he ascended '*Ad astra per aspera*'" (Through hardships to the stars). Lowell, the Martian emissary, had been laid out in his casket in academic regalia and was interred in Arizona soil. Seven years later he would be exhumed when Constance Lowell had her husband's body placed in a structure more befitting a man of his station and ego—a granite mausoleum, domed like an observatory and carved with his own defiant words:

> [AN ASTRONOMER] MUST LEARN TO WAIT UPON HIS OPPORTUNITIES AND THEN NO LESS TO WAIT FOR MANKIND'S ACCEPTANCE OF HIS RESULTS; FOR IN COMMON WITH MOST EXPLORERS HE WILL ENCOUNTER ON HIS RETURN THAT FINAL PENALTY OF PENETRATION, THE CERTAINTY AT FIRST OF BEING DISBELIEVED.

...

MANY YEARS EARLIER, WHEN Lowell was a young man traveling Europe after his graduation from Harvard, he had written a schoolmate back home, "I should rather like to read my own obituary and find out how many good qualities my friends would endow me with when dead which they would be very careful to keep to themselves during my lifetime." Now, upon his death, Lowell's friends offered a mixed assessment of his legacy.

"He got neither the fun nor the glory out of life," one of his old college chums remarked sorrowfully of the great potential that went unfulfilled. Garrett P. Serviss praised the late astronomer's drive and enthusiasm but felt compelled to add, "whether the explanation of [the markings on Mars] offered by Professor Lowell is sound or not is another question." To many scientists, Lowell's story became a cautionary tale, an object lesson in the dangers of allowing emotion to cloud objective judgment. Indeed, his life reads like a cross between a Shakespearean tragedy and a psychological case study, that of a man destroyed by his own self-delusion. Despite his aristocratic airs and his affinity for the superhuman Martians, Lowell possessed, in epic proportions, a mortal flaw that afflicts us all.

Four centuries ago, the English philosopher Francis Bacon described the psychological tendency known today as confirmation bias. "The human understanding when it has once adopted an opinion . . . draws all things else to support and agree with it," he explained, while everything to the contrary "it either neglects and despises, or else by some distinction sets aside and rejects." Percival Lowell had done just that for a decade and a half—scoured the world for justification of his precious theory and downplayed or ignored everything contrary to it. Given the power of his intellect, money, and elocution, he had then been able to persuade the masses to see things as he did, at least for a time. This deception, unintentional though it may have been, spread easily through the sensation-seeking press to a broader, often less educated public in search of meaning in the universe. People in great profusion then, as now, ached to believe in the existence of higher beings, and the fact that Lowell's Martians built canals—at a time when humans were constructing Earth's most ambitious canal—seems hardly a coincidence. Part of what made the Martians so appealing was their relatability. They were us, only better—wiser, more peaceful, more moral.

Although Lowell had sent the world on an excursion into fantasy, such a detour was not necessarily a failing. Science eventually self-corrects, and Lowell's vigorous advocacy pushed other astronomers to refine their own ideas and evidence. The edifice of science rests on the remains of countless discarded theories; error forms an inevitable part of the constructive process. Lowell's fundamental sin was not his misinterpretation of the data or his grand theorizing but rather his dogmatism and obstinacy, his inability to jettison outmoded ideas when they had been shown untenable.

Yet to focus on Lowell's defects is to undersell his gifts. At the time of his death, the public quickly forgave his flaws while newspapers across the land mourned the passing of such a singular being.

"Whether Lowell has added anything definite to 'science' may be debated," *The Hartford Courant* conceded. "But it is certain that he aroused the public interest in the heavens and drew attention to the mysteries of life, widening the public vision, and deepening the sense of the vastness of the universe. Who calls that wasted labor?" The *Cleveland Plain Dealer* offered more unabashed praise: "His work is a great and lasting one, for it has opened amazing fields of speculation, and has led the human imagination to soar to distances seldom before attained." A New Jersey paper headlined its eulogy THE MARS MAN. "Professor Lowell gave untrained men the freedom of the skies, turned the imagination of nations starward and enlarged our conception of what the life of the universe may be," the piece concluded. "This is really a greater achievement, though a poetic achievement, than the exact discovery of what the . . . markings on Mars indicate."

Asked to explain the source of her late husband's enigmatic vision, Constance Lowell struggled for words and reached for Latin. "*Poeta nascitur, non fit*," she wrote. A poet is born, not made.

Another woman, who under different circumstances might have become Percival's wife, offered her own personal remembrance. Thirteen years before Lowell's death, in 1903, Irva Struthers had been wooed by him in Maine and Philadelphia, had been captivated by his stories of Mars and the cosmos. Now, having settled down as Mrs. William W. McCall, she looked back wistfully on her brief and magical relationship with America's poet astronomer. "The Stars have claimed you for their own, my Martian," she recorded in a private tribute. "Sometime will you look down from your bright planet, and with your clearer vision read my heart. Your name is written there." Although she had rejected her suitor as an impractical choice of spouse, she could not deny his magnetism, the force of his ideas. "I did not know until you had gone, the hold you had upon my life, my thoughts, my actions." She had kept his love letters all those years.

Epilogue

Children of Mars

Even after Percival Lowell exited the stage in 1916, the supporting cast in Earth's Martian drama played on for years as each of the storylines drew toward an inexorable, sometimes tragic, conclusion.

Nikola Tesla, his finances in ruins, suffered the indignity of seeing his great communications tower on Long Island dynamited for scrap in 1917, and the destruction of his dream destroyed the man. Unable to afford the high living of the Waldorf-Astoria, he drifted from hotel to hotel, leaving behind a string of unpaid bills as he withered into a recluse whose closest friends were the pigeons he fed in New York's Bryant Park and the aliens with whom he thought he had established contact. "I am expecting to [write] the Institute of France . . . and claim the Pierre Guzman prize of 100,000 francs for means of communication with other worlds," he announced on his eighty-first birthday, in 1937. "I am just as sure that prize will be awarded to me as if I already had it in my pocket." He died, still in debt, six years later.

David Todd's grip on reality grew similarly tenuous. In 1917, Amherst College pressed him into early retirement for exhibiting strangely erratic conduct, and Mabel Todd also noticed her husband's odd behavior: his profligate spending, his tendency to vanish for days at a time, and his renewed fixation on Mars. When doctors pronounced him suffering from "circular insanity," she had him institutionalized.

"WHY DO YOU KEEP ME locked up in a Bug-House against my will . . . ?" he scrawled in a desperate plea to his wife as she unburdened in her journal, "I can't write of it—it is too heart breaking." For two decades, David Todd rolled in and out of mental hospitals, but even that did not keep him from Mars and the media spotlight. He regaled the press with new schemes for studying the planet—using high-flying balloons and an underground telescope—and in 1924, when Mars made its closest approach to Earth in more than a century, he convinced the U.S. military to silence its radios to listen for signals. None were heard.

Camille Flammarion, the starry idealist who long argued that contemplating the heavens would uplift the Earth, found his optimism crushed by Europe's Great War. "Our world is radically imperfect," he despaired in 1917. "Therefore, let's not delude ourselves about the imminent enlightenment of human reason by astronomy." Like Lowell, however, Flammarion never quit the Martians. "Rest assured, we shall some day correspond with the inhabitants of Mars," he vowed in 1923, "not with flying machines but with psychic waves in which I firmly believe. Telepathy will overcome space."

Lilian Whiting retained her hopeful outlook through the tumultuous years of World War I and, again, when World War II visited destruction on Europe. In 1940, certain of a better life in the hereafter, she proclaimed that "a new, a more glorious, resurrection awaits the future of all who keep heart" and urged, "Let us look upward, not downward." By *upward* she may well have meant Mars, for she still found Lowell's vision of the planet to be credible. "Perhaps it is to some such place that we go after death," she once said, imagining her rebirth on that wise and ancient orb.

An aging H. G. Wells, still prolific and influential, railed against fascism and Nazism and warned of dystopian futures, but his best-known books were those of the past. In 1937, in a new and soon forgotten novel that revisited the possibility of life in space, he mocked his famous story of forty years earlier. "Some of you may have read a book called *The War of the Worlds*—I forget who wrote it—Jules Verne, Conan Doyle, one of those fellows," remarked a passing character who dismissed the ridiculous improbability of a Martian invasion. Yet the following year, when Orson Welles adapted *The War of the Worlds* for radio, the premise struck some in the audience as all too real. One duped listener explained of her gullibility, "Has not science proved that there is life on Mars, through their observations? Canals are supposed to have been discerned which prove that there is the possibility of their being on a higher level than ours in scientific development." Twenty years had passed since Lowell's death, and yet his ghost lingered.

Lowell's specter haunted the scientific realm too. "Now the story of the 'canals' is a long and sad one, fraught with back-bitings and slanders; and many would have preferred that the whole theory of them had never been invented," a British astronomer wrote in the 1930s. A leading American astronomer said much the same *in the 1960s*. "This controversy brought disrepute to planetary science and weakened its status in universities," Gerard Kuiper contended. "To this day the effects have not been overcome." And yet, whatever injury Lowell may have caused the progress of science, he simultaneously did much to drive it forward.

...

GARRETT P. SERVISS, WHO had played such a central role in publicizing the Martian canal theory, confronted its mixed legacy as he neared the end of his own life in the 1920s. After he gave a lecture one day, a small girl approached. Her head reached no higher than the band on his wristwatch. "Are there people on Mars?" she asked.

The wizened science journalist looked down at his young interrogator. "My dear, I don't know. I wish there were people there, but— I'm afraid there are none." He then asked, "Do you believe there are people on Mars?"

"Yes!" the girl insisted.

Seeing the passion in her eyes and not wanting to extinguish it, Serviss then amended his answer to allow at least the faint possibility of intelligent beings on Mars, for who could prove that the Martians were *not* there? He learned a lesson from the encounter: Even in science, one should leave room for hope and imagination. "It was a compliment to astronomy that that child should have been reading about Mars," Serviss concluded. "That dream of hers was a thing to be encouraged instead of being destroyed."

As Serviss discovered, during the period of the Mars craze, intelligent eyes really had been watching human affairs keenly and closely, like the Martians of *The War of the Worlds*. These advanced beings, however, did not gaze down from the heavens but rather up from below. Children, hiding in the interstices of society, had been observing the Mars excitement more closely than adults appreciated, absorbing ideas that would transform the world when the young, no longer young, matured into its leaders.

In 1898, when the *Boston Post* serialized *The War of the Worlds* and *Edison's Conquest of Mars*, a sickly teen with a penchant for things mechanical lost himself in the thrilling narrative day after day. On a quiet October afternoon the following year, as he climbed a cherry tree behind his grandmother's barn in Central Massachusetts, a vision came to him from those stories. "I imagined how wonderful it would be to make some device which had even the *possibility* of ascending to Mars, and how it would look on a small scale, if sent up from the meadow at my feet," Robert H. Goddard later recalled. "I was a different boy when I descended the tree." His life's goal became to design an actual spaceship. Some three decades later, Goddard would construct the world's first liquid-fueled rocket, launching the space age.

Meanwhile, in 1890s Luxembourg, a precocious youth read Lowell's *Mars* and grew so transfixed that he reportedly "lapsed into delirium, raving about strange creatures, fantastic cities, and masterly engineered canals of Mars for two full days and nights." His name was Hugo Gernsback, and after moving to the United States as a young adult, he turned his childhood passion into a publishing phenomenon. Gernsback established a series of magazines that printed futuristic stories, often about Mars and space travel, a genre he called "scientifiction" before rebranding it into the literary phenomenon it would become. A trailblazing editor now widely revered as the "father of science fiction," Gernsback remains memorialized in the annual sci-fi literature awards named in his honor, the Hugos.

Many twentieth-century writers of science fiction also traced their inspiration to childhood fervor about Mars. Robert S. Richardson, a professional astronomer who penned stories under the name Philip Latham, pointed to his early discovery of Lowell's *Mars and Its Canals*. "I was about twelve then, and the book made a profound impression upon me. I believed it absolutely. In fact, it was inconceivable to me that anyone could *not* believe it." For Edmond Hamilton, a pioneer of the subgenre known as space opera, the spark came at age four when he stumbled on H. G. Wells's winged, angelic Martians in *Cosmopolitan Magazine*. "I looked at that magazine until it wore out," Hamilton recalled. "I wasn't yet able to read it, to read the article, but those pictures! I sat and wondered if Mars was a long way off and if it was a very strange place." H. P. Lovecraft, remembered as a master of fantasy and the macabre whose stories featured extraterrestrials, started as a boy infatuated with astronomy. He attended one of Lowell's lectures when just sixteen and declared the theory of an inhabited Mars "not only possible, but even probable."

Lowell's influence leapfrogged to a whole new generation when the creator of another craze—the Tarzan novels—wrote a string of adventure books set on a fictional Mars known by its inhabitants as Barsoom. Home to ancient cities, dead sea bottoms, and a global network of irrigation canals, here was Lowell's dying planet vividly realized, with one crucial difference: rather than a world at peace, Barsoom had devolved into feudal warfare. *A Princess of Mars*, the book that inaugurated the series, introduced a daring and chivalrous earthling named John Carter, who transported himself through space simply by wishing it so with arms outstretched. Once on Mars this interplanetary knight battled his way through hostile tribes to rescue the handsome princess of Helium, Dejah Thoris. Author Edgar Rice Burroughs spun the tale in such muscular prose that he fired the imagination of millions of adolescent boys and not a few girls.

Among the ardent young readers of the Barsoom novels were future writers, including Ray Bradbury, Arthur C. Clarke, and Leigh Brackett, who would set their own science-fiction tales on Mars and pass Lowell's imaginative torch on to yet another generation.

A future scientist, too, spent his early years preoccupied with John Carter's space adventures and made his own attempts to emulate them. "I can remember spending many an hour in my boyhood, arms resolutely outstretched in an empty field, imploring what I believed to be Mars to transport me there," Carl Sagan reminisced. "It never worked. There had to be some other way."

...

THE FAIRY TALES OF youth are not forgotten; their imprint remains chiseled and sculpted into the beings we become. What is true for the individual applies equally to society. The stories we once told ourselves as a civilization, our legends about dragons and demons and Martians, inhabit our culture and collective unconscious. "The fact is, that ideas are as catching as scarlet fever," Lowell himself once wrote. "We can no more escape having them enter our minds than we can escape having material germs enter our bodies." His ideas about Mars, although they had moved from reality to myth, infected one generation after another until they eventually pushed us into space to learn the truth about our neighboring world.

On Friday, November 12, 1971—autumn in California, late spring in the Southern Hemisphere of Mars—an octagonal metal box sporting solar panels and electronic eyes was about to become the first spacecraft from Earth to orbit another planet. *Mariner 9* would spend a year swooping toward the Martian surface to take thousands of photographs in unprecedented detail, and at the Jet Propulsion Laboratory in Pasadena, a clutch of noted scientists and science writers had assembled for this historic moment. It was a celebration tinged with a sense of loss, for whatever the probe was about to find, it would almost certainly destroy the vision that had so captivated humanity for the better part of a century.

As if to eulogize a planet that never was, some of the big names in attendance held a public discussion to reflect on Percival Lowell's impact and what his Martians revealed about the human mind. "Whatever we can say about Lowell's observational abilities, we can't deny his propagandistic power," said Arthur C. Clarke, who had memorialized the early canalists in his novel *The Sands of Mars* when he named two Martian towns Port Lowell and Port Schiaparelli. "[Lowell] certainly did a lot of harm in some ways, but I think perhaps in the long run the benefits may be greater."

Ray Bradbury, whose literary stardom launched with *The Martian Chronicles*, agreed: "I think it's part of the nature of man to start with romance and build to a reality. There's hardly a scientist or an astronaut I've met who wasn't beholden to some romantic before him who led him to doing something in life."

Joining the conversation was the astronomer and science popularizer Carl Sagan, who, as a member of the *Mariner 9* imaging team, had finally found his way to Barsoom, for he would be among the first to see the pictures beamed back of the Martian surface. "I don't think there is an advanced civilization on Mars," he acknowledged. "But we cannot exclude it. The remarkable thing is that *Mariner 9* is the first mission which gives us the capability to test such a hypothesis."

In the following months, Sagan scrutinized pictures of the Martian landscape for signs of life. He observed things that Lowell had seen: markings that darkened and lightened throughout the year, as if crops were growing and decaying with the seasons, but Sagan concluded these were mere sand paintings—shifting patterns of bright, windblown dust that hid and then revealed dusky underlying rock. As for the canals, Sagan examined Lowell's old maps to identify what features on the planet might explain the placement of the cobwebbed lines. In a few cases, Lowell's canals matched up with linear rift valleys or ridge systems, but the vast majority correlated with nothing on the Martian surface. The lines had been mere chimeras, an amalgam of misperceptions due to atmospheric distortion, the fallible human eye, and one man's unconstrained imagination.

"Even if all Lowell's conclusions about Mars, including the existence of the fabled canals, turned out to be bankrupt," Sagan wrote forgivingly, "his depiction of the planet had at least this virtue: it aroused generations of eight-year-olds, myself among them, to consider the exploration of the planets as a real possibility, to wonder if we ourselves might one day voyage to Mars."

EPILOGUE

...

AS A SPECIES, WE have now voyaged to Mars more than two dozen times, on the backs of spacecraft and robotic rovers with names hopeful and heroic: *Viking, Pathfinder, Spirit, Opportunity*. We have traveled the rocky mesas and ridges, the broad plains of sand and gravel, the desert scoured by a phantom wind. Along the way we have learned far more about the planet than Lowell and his contemporaries could ever have dreamed.

We now know that Mars is a frozen world in more than one sense, so cold that the very air crystallizes—forming dry ice at the poles (where Lowell had imagined great stores of water)—and where geological processes have largely halted. Mars really is a dying planet, having lost its magnetic field and most of its atmosphere. The rocks and landforms suggest, however, that in its infancy Mars may have been warm and earthlike, a place of spewing volcanoes, surging rivers, and tranquil lakes. Young Mars could well have been a nursery for life.

Today, many scientists imagine Mars as a place where simple organisms arose several billion years ago and where the fossilized remains—indeed, the living descendants—might still be underground, shielded from the radiation that now sterilizes the Martian surface. To test that theory, NASA's *Perseverance* rover has drilled rock cores on Mars and sealed them in tubes for eventual transport to Earth, where geochemists hope one day to examine the samples for signs of life.

Other Martian rock samples have already arrived on Earth, on their own. In 1984, researchers discovered a meteorite in Antarctica that they determined to be a small chunk of Mars, blasted off the surface by an impact there that launched it on a trajectory here. That meteorite was later found to contain hints of organic chemistry and what looked like fossil bacteria, spurring a modern-day Mars "boom" when President Bill Clinton hailed the news as the possible first discovery of alien life. Most scientists now reject that interpretation, but the Antarctic find bolstered an even more radical idea—that if life *did* originate on Mars, it may have seeded life on Earth by just such an interplanetary exchange of rocks. In other words, we may all be of Martian birth deep in our evolutionary past.

As for humankind's future, today's Mars enthusiasts consider our neighboring world to be civilization's destiny. They foresee what is now an inhospitable wasteland eventually harboring temporary settlements, then permanent cities. In fact, some suggest that people might one day reengineer the entire globe, restoring its atmosphere and water cycle, making Mars a second Earth. These ideas may sound outlandish, and in a hundred years they could well be dismissed as the naïve beliefs of a backward people, much as Lowell's Martians are now viewed. Yet I no longer scoff at such improbable Mars dreams.

When I began this book, I thought I had set out to tell a story of human folly, about how easy it is to deceive ourselves into believing things simply because we wish them to be true. That is, of course, one lesson of the tale, but I discovered another, perhaps more powerful, takeaway: Human imagination is a force so potent that it can *change* what is true. Thanks to Lowell's Martian fantasies that helped inspire the early space age, visiting the red planet has become a potentially realizable goal for today's children. Both NASA and the Chinese space agency hope to begin sending astronauts to Mars as soon as the 2030s, and a conspicuous tech billionaire with his own rocket company aims to make interplanetary travel so routine and inexpensive that most anyone could emigrate to Mars for less than the cost of a middle-class home in America.

To cast such airy notions into concrete reality, an advocacy group called the Mars Society was founded in 1998 under a bold declaration that begins, "The time has come." This heady band of zealots recently held a design competition, a request for proposals for the first Martian city of a million souls. The contest drew almost two hundred entries from across the globe, and to read them is to see how little our Martian imaginings have changed in a century. An American design depicts the inhabitants living in circular, domed communities—oases, if you will, filled with museums and parks and fine dining—connected one to another via straight-line conduits, a geometry that mimics Lowell's canal network. Even more Lowellian is a British proposal that calls for a city "laid out in a spider-web shape." To slake the thirst of the residents and to grow crops, almost all the proposed cities rely on meltwater, although not from the seasonal thawing of the polar caps but from the mining of ice trapped in craters or underground.

When I read these design proposals, what most struck me was not their technical aspects but their portrayal of Martian culture. The plans describe cities filled with a multicultural mix of artists and engineers who celebrate festivals, play low-gravity sports, live in harmony, and recycle their dead. Healthcare is free, as is education, and all workers earn the same pay, since every job is essential to the colony's well-being. As one team of Martian urban planners put it, to survive in such a harsh environment, the inhabitants must create a federation "based on solidarity, benevolence, generosity, equality and kindness."

A century ago, such visions of humanity perfected on Mars may have been fables, but dreams are more than delusions. They expose our deep-set longings, our hungers unmet. Today's Martians are still what they were in Lowell's time: the embodiment of hope for a better world—if not on Earth, then not so far away, on the next planet outward from the sun.

Acknowledgments

In the fall of 2017, fresh off the publication of my book *American Eclipse* and looking for a suitable new topic to write about, I recalled a whisper of memory from childhood. Sometime in early life—I forgot when and how—I had learned the outlines of Percival Lowell's story: that an educated and erudite man had once seen lines on Mars, had interpreted them as signs of intelligence, and had fooled himself and a good portion of the public into believing in the existence of Martians. Passionate about Mars myself, I wondered if this might provide the kernel of a new book.

I started by digging into newspapers from the early twentieth century, to see how Lowell's theories had been interpreted during his lifetime. What I found astonished me. The tale of this man and his far-reaching ideas seemed mythic, rich with lessons about human longings and failings, so I brought these discoveries to my gifted literary agent, Todd Shuster. Together with his talented associates, Justin Brouckaert and Jack Haug, Todd deftly steered me on the right course as I launched into this project. He then reconnected me with my inspiring editor, Bob Weil, who shaped my thinking and improved my manuscript in profound ways. I am tremendously grateful to Bob and to his crackerjack assistant, Luke Swann, who offered additional editorial guidance and cheerfully shepherded my book through the production process.

I could not have completed a book of this scope without the support of many people and institutions. Beginning with the latter, I am grateful to Arizona State University for granting me a journalism fellowship in transdisciplinary science, an intellectually stimulating month at ASU's School of Earth and Space Exploration during which I met distinguished physicists, astronomers, geologists, and biologists who are exploring Mars and imagining life in space. These conversations in early 2020 (just before the pandemic brought the world to a halt) provided me with the scientific context I needed to put my historical tale in perspective. The following year, the John W. Kluge Center at the Library of Congress selected me as its Baruch S. Blumberg NASA/Library of Congress Chair in Astrobiology, Exploration, and Scientific Innovation. This appointment placed the enormous resources of America's national library at my disposal, enabling me to peruse collections of rare material—published and unpublished—as I pieced my story together. I am indebted, as well, to the Alfred P. Sloan Foundation, which underwrote further research and travel—to Arizona, Massachusetts, Connecticut, France, Italy, Chile—to uncover documents and retrace the footsteps of my characters. Closer to home, the University of Colorado's Center for Environmental Journalism kindly appointed me a visiting scholar, a status that conferred valuable borrowing privileges from CU's libraries. To say that I made good use of those privileges would be an understatement. My accumulation of simultaneously checked-out items peaked at 192 library books on my home shelves in the fall of 2022.

Of the many people who assisted me in my research and writing, several deserve special mention. In the winter of 2018, while appearing as a guest speaker in the Pathway to Space course on the CU campus, I had the good fortune to meet Professor Bruce Jakosky. A distinguished planetary scientist who spearheaded NASA's MAVEN mission to Mars, he invited me to lunch to discuss my nascent book, and thus began an enduring friendship. During my years of work, Bruce generously shared his time, passion, and expertise, motivating my efforts and supporting them in practical ways. Over the same period, Lauren Amundson, archivist at the Lowell Observatory, served as my guide to that institution's valuable collections. She welcomed me during my research visits to Flagstaff, answered my endless queries, and—after I returned home—graciously scanned documents I had neglected to photograph while on-site. William Sheehan, a prolific historian of astronomy who has written brilliantly about Lowell and Mars (and is now a friend whom I call Bill), helped in so many ways—sending me his works-in-progress, pointing me toward relevant documents, and introducing me to his astronomical contacts around the world.

Among those to whom Bill introduced me was Agnese Mandrino, archivist at the Brera Observatory, who—with great energy and efficiency—led me through Giovanni Schiaparelli's logbooks and correspondence when I visited Milan. While there, I also benefited from the generous help of Mario Carpino, who took me to the rooftop telescope through which Schiaparelli first spied the Martian *canali* in 1877. In France, I enjoyed the hospitality of Charles White and Francis Oger, who escorted me through the library, château, and grounds of Camille Flammarion's observatory in Juvisy. Thanks to David Valls-Gabaud for arranging that visit and for helping me track down material in the Juvisy archives. When I later planned a trip to Chile, Louise Purbrick and Xavier Ribas kindly advised me on how to locate the ruins of Oficina Alianza in the vast Atacama. At the Lowell Observatory, historian Kevin Schindler generously acted as tour guide to Mars Hill, and Kevin White and Liza Matrecito served as telescope operators for the thrilling hour I spent observing the red planet through Percival Lowell's great lens. Lowell's living relatives also assisted me. I am grateful to Lowell Putnam and Carol Bundy for alerting me to the trove of unprocessed family papers in the collections of the Massachusetts Historical Society, and to Jennifer Putnam for kindly sharing her expert translations of the correspondence between Lowell and Schiaparelli.

For helping me track down additional source material, I thank eminent historians of astronomy: Louise Devoy, David DeVorkin, Arnold Heiser, David Strauss, Richard McKim, and Sara Schechner. For sharing history of a more personal nature, I am grateful to Michelle Koross, Herb Koross, Elinor Hipscher Koll, and Bernie Mansbach. They recounted family memories of the 1938 radio broadcast of *The War of the Worlds*, stories that played a prominent role in an early draft of this book. I apologize to them for cutting those colorful tales when they no longer fit my manuscript's restructured narrative.

A dedicated group of friends and colleagues donated their time and talents to reading and commenting on my manuscript. I am hugely grateful to Kathryn Bowers, Steven Dick, Rachel Nowak, Len Ackland, Dan Glick, and Bill Raymond, as well as the aforementioned Bruce Jakosky and Bill Sheehan. Others who offered helpful ideas and stimulating conversation during my years of Mars obsession include Andrea and Dana Meyer, Tom English, Deborah Kent, Steve Hayward, Jay Pasachoff, Scott Carney, Tom Yulsman, Erin Espelie, and Cathy Koczela. Thanks to Alison Perlo for correcting and improving my translations from French to English, and to Michael Saks and Roselle Wissler for showering me with Arizona hospitality during my visits to Phoenix and Flagstaff. My gratitude goes, as well, to copyeditor Rachelle Mandik, who polished my manuscript to an expert shine.

Other individuals I must recognize include the ASU faculty members and staff who helped me during my fellowship in Tempe: Skip Derra, Nikai Salcido, Karin Valentine, Mini Wadhwa, Phil Christensen, Jim Bell, Paul Davies, Lindy Elkins-Tanton, Evgenya Shkolnik, Lance Gharavi, Cady Coleman, Andrew Maynard, Ariel Anbar, Timiebi Aganaba, Steve Ruff, Sara Walker, Steve Desch, Hilairy Hartnett, Dave Williams, Michael Varnum, Ramon Arrowsmith, Patrick Young, Don Burt, Everett Shock, and Jack Farmer. At the Library of Congress, I had the good fortune to work with Dan Turello, Travis Hensley, Michael Stratmoen, John Haskell, Andrew Breiner, David Konteh, Josh Levy, Gary Johnson, Domenic Sergi, Michael Neubert, and Nida Khan. I also thank Lucas Mix, my successor as Blumberg Chair, whose appointment overlapped with mine. Our weekly lunches at Pete's Diner were a highlight of my term in Washington, D.C.

I am blessed to have a family that supports and encourages me through my strange obsessions and long writing projects. I am grateful to all, most especially my husband, Paul Myers. During solitary weeks, when I would emerge bleary from hours spent in the company of century-old newspapers, it was Paul with whom I shared my latest discoveries and setbacks, and he was the first to review my chapters when they were finally worth reading. Like so much in life, writing is more about the journey than the destination. On the long trek that was this book, I could not have asked for a better traveling companion.

As I entered the final stages of crafting this book, in the autumn of 2023, I made an unexpected find. To assist my writing of the epilogue, I rewatched Carl Sagan's *Cosmos*, a television series that left a huge mark on my youth when it debuted on PBS in 1980. Back then, at age sixteen, I had found myself mesmerized by this eloquent astronomer who took me on his spaceship of the imagination to tour the wonders of the universe, an excursion that inspired my own career as a science communicator. Viewing the series again at fifty-nine, my eyes widened when I saw Sagan at Percival Lowell's observatory in Flagstaff.

"It was here that the most elaborate claims in support of life on Mars were developed," Sagan told his audience in episode 5. "Lowell believed that he was seeing a globe-girdling network of great irrigation canals carrying water from the melting polar caps to the thirsty inhabitants of the equatorial cities. . . . He believed that the planet was earthlike. All in all, he believed too much." So this was how I had first learned of Percival Lowell and the canals of Mars—from Carl Sagan. I should have known.

Sagan became a planetary scientist out of his boyhood fascination with the Mars novels of Edgar Rice Burroughs, and Burroughs had been inspired by the theories of Percival Lowell. I must therefore thank Lowell himself for planting the seed that would eventually germinate into this book. Not only did Lowell—that peculiar, creative, stubborn Bostonian—provide the subject matter for my story, but he gave it the imaginative impulse. He lit the fire of Mars passion that has burned through the ages, from generation to generation and brain to brain—from Lowell to Burroughs, Burroughs to Sagan, Sagan to me.

Notes on Sources

Writing this book proved a far more challenging project than I had originally imagined. The deeper I dug into the events and preoccupations of the turn of the twentieth century, the more I discovered interlinking strands that connected back to Mars, much like the web of the fabled canals themselves. I found myself researching Darwinian evolution, the yellow press, the Spanish–American War, the development of wireless communications, the St. Louis World's Fair, the Panama Canal, the Wright brothers, the race for the North Pole, and so much more. Meanwhile, as I explored the historical context for my story, I had to reconstruct the central tale itself, which required that I delve into the minute details of my many characters. I could not have untangled this complex network of people and ideas without the insights provided by innumerable authors before me. Please see my bibliography for a list of the most important published sources I consulted. Here I shall highlight some of them.

My starting point for researching Lowell and Mars was, unsurprisingly, *Lowell and Mars*, the classic account by William Graves Hoyt. I followed up with David Strauss's intellectual biography *Percival Lowell*, as well as the more hagiographic *Biography of Percival Lowell*, written by Percival's brother, A. Lawrence Lowell. Anyone interested in learning more broadly about the remarkable Lowell family and its influence would do well to read Nina Sankovitch's exquisite *The Lowells of Massachusetts*.

When writing about other characters in this drama, I benefited from many biographical books and articles: *Edward Sylvester Morse*, by Dorothy Wayman; *Camille Flammarion*, by Philippe de la Cotardière and Patrick Fuentes; "The Life and Times of E. M. Antoniadi, 1870–1944," by Richard McKim; and *Tree Rings and Telescopes*, George Ernest Webb's account of the life and career of A. E. Douglass. I consulted numerous books about H. G. Wells, including those by David C. Smith, Anthony West, Claire Tomalin, and Wells himself. (See my bibliography for titles and details of all these works.) Of the multiple biographies of Nikola Tesla I read, I found W. Bernard Carlson's excellent *Tesla* especially useful and trustworthy. As for David Todd, his life has not been chronicled in a book, but several books focus on his multitalented wife. Mabel Loomis Todd's role in publishing the poetry of Emily Dickinson is wonderfully described in *After Emily*, by Julie Dobrow. Todd's passionate romance with Emily Dickinson's brother sears the pages of *Austin and Mabel*, by Polly Longsworth.

As I researched humankind's changing beliefs about aliens, I leaned heavily on two masterful works: Michael J. Crowe's *The Extraterrestrial Life Debate, 1750–1900*, and Steven J. Dick's *Plurality of Worlds*. To understand the history of the scientific study of Mars, I relied on the unparalleled scholarship of William Sheehan, who has produced a deluge of superb articles and books, beginning in 1988 with *Planets and Perception*. (Again, see bibliography.) To trace shifting ideas about Mars in popular culture, I consulted several penetrating volumes of recent years: *Dying Planet*, by Robert Markley; *Imagining Mars*, by Robert Crossley; *Geographies of Mars*, by K. Maria D. Lane; and *News from Mars*, by Joshua Nall. I also highly recommend *The Sirens of Mars*, an evocative mix of memoir and natural history by the planetary scientist Sarah Stewart Johnson.

Nothing beats the sheer virtuosity of Barbara Tuchman's *The Proud Tower* for revivifying life in Europe and the United States prior to World War I. Other books that expertly evoke the era, its obsessions, and blind spots are *No Place of Grace*, by T. J. Jackson Lears; *1898*, by David Traxel; *The Great Wave*, by Christopher Benfey; *To Conquer the Air*, by James Tobin; and—though a work of historical fiction—E. L. Doctorow's epic *Ragtime*. Much has been written about the North Pole controversy of 1909. I highly recommend *The Noose of Laurels*, by Wally Herbert, and *Battle of Ink and Ice*, by Darrell Hartman. Hartman's book also provides an excellent introduction to the rough-and-tumble New York press of the era, a topic explored more fully in Allen Churchill's *Park Row*, David Nasaw's *The Chief*, and Joyce Milton's *The Yellow Kids*. An even better way to get a taste of journalism in that era is to read the actual newspapers, a task easily accomplished today via the many archival news databases online. Not all publications are available in digitized form, however. Of necessity, I conducted some of my newspaper research by means of old-fashioned microfilm and, even more antiquated, by reading original bound volumes.

All of the sources mentioned above—and many others—set me on the path to writing this book, but the story really came alive when I explored the personal papers of the individuals whose tales I tell. Manuscript collections in the United States and Europe contain the correspondence of Percival Lowell, Giovanni Schiaparelli, Camille Flammarion, David Todd, Mabel Loomis Todd, Edward S. Morse, Edward Holden, Nikola Tesla, H. G. Wells, E. Walter Maunder, E. M. Antoniadi, A. E. Douglass, and others. Garrett P. Serviss, unfortunately, left behind no storehouse of personal papers, but his prolific output in newspapers and magazines allowed me to trace his movements and intellectual development. In similar manner, Lilian Whiting's weekly columns in the Chicago *Inter Ocean* and the New Orleans *Times-Democrat* served as a kind of public diary, revealing her thoughts and passions as they evolved over the years.

Most of my citations in the notes that follow are to letters, diaries, and articles that were written at the time of the events recounted. News items in this era were often syndicated or brazenly copied by other papers, which means that many published quotations appeared in multiple places. I have tried, when possible, to cite the original source rather than a reprint. I do not offer references for facts easily ascertained from obvious sources, such as the details of prominent events in history or in the lives of the famous. The reader may also safely assume that my descriptions of weather conditions come from contemporaneous newspaper reports.

When quoting from published articles and archival documents, I have preserved the original spelling and punctuation, with a few exceptions: Where a handwritten item employs underscoring for emphasis, I have generally changed this to italics, and if a quotation represents what someone *said* rather than *wrote* (for instance, Percival Lowell's lectures or David Todd's comments to reporters), I have updated archaic spelling and punctuation to reflect how these spoken words would be transcribed today.

I am careful to refer to newspapers by their official titles. For instance, it was *The New York Herald* ("the" was part of its name), the *New York Evening Journal* ("the" was *not* part of its name), and the New York Sun (in other words, *The Sun*, published in New York). I do veer from this rule in a few cases, however. William Randolph Hearst's morning New York paper went by many titles over a short span of years: *The Journal*, the *New York Journal*, the *New York Journal and Advertiser*, the *New York Journal and American*. For ease of storytelling, I simply call the paper the *New York Journal* in the body of this book. Similarly, although *The New York Times* was called *The New-York Times* until 1896, I omit the hyphen when discussing the paper in this book's narrative so as not to confuse or distract. In the notes that follow, however, I always cite a paper's title as it stood at the time.

ARCHIVAL ABBREVIATIONS

The following abbreviations are used for cited archival collections:

AEDP – Andrew Ellicott Douglass Papers, Special Collections, University of Arizona Libraries, Tucson

AGBFP – Alexander Graham Bell Family Papers, Manuscript Division, Library of Congress, Washington, D.C.

ALC – Amy Lowell Correspondence (MS Lowell 19-19.4), Houghton Library, Harvard University, Cambridge, Massachusetts

ALRP – Abbott Lawrence Rotch Papers, Manuscript Division, Library of Congress, Washington, D.C.

APCR – American Play Company Records (*T-Mss 1966-002), Billy Rose Theatre Division, New York Public Library for the Performing Arts, New York

CFP – Camille Flammarion Papers, Archives of the Société Astronomique de France, Juvisy, France

DPTP – David Peck Todd Papers (MS 496B), Manuscripts and Archives, Yale University Library, New Haven, Connecticut

ECSP – E. C. Slipher Papers, Lowell Observatory Archives, Flagstaff, Arizona

NOTES ON SOURCES

EEBP – Edward Emerson Barnard Papers, Special Collections and University Archives, Vanderbilt University, Nashville, Tennessee
ESMP – Edward Sylvester Morse Papers, Phillips Library, Peabody Essex Museum, Rowley, Massachusetts
FJSL – Frederic Jesup Stimson Letters from Various Correspondents (MS Am 1681), Houghton Library, Harvard University, Cambridge, Massachusetts
FSDG – Fondo Schiaparelli, Domus Galilaeana, Pisa, Italy
FVGP – Francis Vinton Greene Papers, Manuscripts and Archives Division, New York Public Library, New York
GBAP – Papers of George Biddell Airy, Royal Greenwich Observatory Archives, Cambridge University Library, Cambridge, U.K.
GCFP – George Clayton Foulk Papers, Manuscript Division, Library of Congress, Washington, D.C.
GEHP – George Ellery Hale Papers, Archives and Special Collections, California Institute of Technology, Pasadena
GIHC – General Information about Harvard Commencement and Class Day in 1874, 1875, 1876, Harvard University Archives, Cambridge, Massachusetts
GSP – Giovanni Schiaparelli Papers, Brera Observatory Archives, Milan, Italy
HFP – Hussey Family Papers, Bentley Historical Library, University of Michigan, Ann Arbor
HLGP – Henry L. Giclas Papers, Lowell Observatory Archives, Flagstaff, Arizona
HLLO – Historic Logbooks, Lowell Observatory Archives, Flagstaff, Arizona
HPLO – Historic Photographs, Lowell Observatory Archives, Flagstaff, Arizona
HUP – Harvard University Archives Photograph Collection: Portraits, Harvard University Archives, Cambridge, Massachusetts
ISGP – Isabella Stewart Gardner Papers, Isabella Stewart Gardner Museum, Boston, Massachusetts
JLGP – John Lowell Gardner Jr. Papers, Isabella Stewart Gardner Museum, Boston, Massachusetts
LELP – Letters to Elizabeth Lowell Putnam (Mrs. William Lowell Putnam) and William Lowell Putnam (MS Am 2078), Houghton Library, Harvard University, Cambridge, Massachusetts
LFP – Lowell Family Papers (MS Am 3166), Houghton Library, Harvard University, Cambridge, Massachusetts
LOAC – Copybooks, Lowell Observatory Archives, Flagstaff, Arizona
LOR – Lick Observatory Records: Correspondence (UA.036.Ser.1), University Archives, University Library, University of California, Santa Cruz
LPFP – Lowell-Putnam Family Papers, Massachusetts Historical Society, Boston [Collection unprocessed as of January 2025]
MLTP – Mabel Loomis Todd Papers (MS 496C), Manuscripts and Archives, Yale University Library, New Haven, Connecticut
MTBP – Millicent Todd Bingham Papers (MS 496D), Manuscripts and Archives, Yale University Library, New Haven, Connecticut

NTC – Nikola Tesla Correspondence, Manuscript Division, Library of Congress, Washington, D.C.
NTM – Nikola Tesla Museum, Belgrade, Serbia
NTP – Nikola Tesla Papers, Rare Book and Manuscript Library, Columbia University, New York
PLD – Percival Lowell Diary: Manuscript, 1904 (MS Am 2018), Houghton Library, Harvard University, Cambridge, Massachusetts
PLP – Percival Lowell Papers, Lowell Observatory Archives, Flagstaff, Arizona
RAAS – Records of the American Astronomical Society, Niels Bohr Library and Archives, American Institute of Physics, College Park, Maryland
RASL – RAS Letters, Archives and Manuscripts, Royal Astronomical Society, London, U.K.
RHCOD – Records of the Harvard College Observatory Director Edward Charles Pickering, Harvard University Archives, Cambridge, Massachusetts
RWOWP – Richard Wilson – Orson Welles Papers, Special Collections Research Center, University of Michigan, Ann Arbor
SNP – Simon Newcomb Papers, Manuscript Division, Library of Congress, Washington, D.C.
TBPC – Todd-Bingham Picture Collection (MS 496E), Manuscripts and Archives, Yale University Library, New Haven, Connecticut
VMSP – V. M. Slipher Papers, Lowell Observatory Archives, Flagstaff, Arizona
WFP – Wendell Family Papers (MS Am 1907–1907.1), Houghton Library, Harvard University, Cambridge, Massachusetts
WHPP – Papers of William Henry Pickering, Harvard University Archives, Cambridge, Massachusetts
WWOWP – Wilbur Wright and Orville Wright Papers, Manuscript Division, Library of Congress, Washington, D.C.

EPIGRAPH

ix **"Mankind has to all intents and purposes"**: P. Lowell (1906:20).

PROLOGUE: WHY MARS?

1 **"Mortals, I am here to tell you"**: *The Washington Post*, February 12, 1907, 12. For background on Mamie "Mrs. Stuyvesant" Fish, see G. King (2009:82–86).
1 **dropped by for New Year's Eve**: *The Washington Post*, January 2, 1907, 7.
1 *A Yankee Circus on Mars*: *The New York Herald*, April 13, 1905, 3–4; *The Sun* (New York), April 13, 1905, 5.
2 **"The Martians may be very fine"**: *Binghamton Press and Leader*, March 9, 1908, 5.
2 **"great centres of population"**: *The New York Times*, December 9, 1906, Part Four, 1.
2 **SCIENTISTS NOW KNOW POSITIVELY:** *The World* (New York), July 8, 1906, Sunday Magazine, 1.
2 **"no escape from the conviction"**: Bell (1910:550). See also Alexander Graham Bell to Mabel Hubbard Bell, November 29, 1909, AGBFP, Box 49 and Box 121.
2 **"*Some day theology*"**: H. Menke to Percival Lowell, May 1, 1907, PLP, Percival Lowell Correspondence.

NOTES ON SOURCES 255

CHAPTER 1: EVOLUTION

10 **Harvard University was holding graduation exercises:** Details of Harvard's 1876 commencement ceremony come from *Boston Evening Journal*, June 28, 1876, 2; *Daily Evening Traveller* (Boston), June 28, 1876, 2; *Boston Daily Advertiser*, June 29, 1876, 2; *The Boston Daily Globe*, June 29, 1876, 1 and 4; *The New-York Times*, June 29, 1876, 2; and *New-York Tribune*, June 29, 1876, 2.
10 **"As a traveler wandering through":** Percival Lowell, "The Nebular Hypothesis," tucked within the volume "Commencement Parts 1874–76," GIHC, Box 1.
11 **"quite profound, but somewhat dry":** *The Boston Daily Globe*, June 29, 1876, 1.
12 **"Everybody nowadays talks":** *The Cornhill Magazine*, January 1888, 34.
12 **convulsed so much of the Christian world:** Bowler (2012). Judaism and Islam proved more accepting of Darwinism; see Cantor (2006) and Iqbal (2009).
12 **evolution implied *progress*:** Lears (1981:20–23).
12 **"a provokingly easy way of dressing":** *The Atlantic Monthly*, January 1860, 92. Also O. W. Holmes (1861:13–14).
13 **sown the seeds of America's Industrial Revolution:** Rosenberg (2011).
13 **"Here things are rather dull":** Percival Lowell to Elizabeth Lowell, May 10, 1882, LELP.
13 **"I have . . . become decidedly misanthropic":** Percival Lowell to Barrett Wendell, January 16 [1877], WFP.
14 **offered by the Lowell Institute:** Everett (1840); H. K. Smith (1898); Greenslet (1946); Weeks (1966).
14 **stout neo-Renaissance building:** This was later known as the Rogers Building. Anderson (1984); Stratton and Mannix (2005:319–339).
14 **"Going to Japan is like visiting a new world":** *Buffalo Commercial Advertiser*, February 25, 1880, 3.
15 **"European dress and ideas":** *Daily Evening Traveller* (Boston), January 10, 1882, 1.
15 **Morse was ambidextrous and artistic:** Wayman (1942:205); *St. Johnsbury Caledonian* (St. Johnsbury, VT), March 31, 1882, 2.
15 **"No other of the several":** *Boston Evening Transcript*, February 18, 1882, 4.
15 **spellbound, was Percival Lowell:** Strauss (2001:122).
15 **"of real significance":** A. L. Lowell (1935:5). See also Yeomans (1948:20–22).
16 **Rose Lee was the daughter:** Sheehan and Bell (2021:141).
16 **Lowell grew disconsolate:** Percival Lowell to Katharine B. Lowell (his mother), July 8, 1882, and A. Lawrence Lowell to Katharine B. Lowell, July 10 [1882], LPFP. Also Percival Lowell to Elizabeth Lowell, July 19 and August 2, 1882, LELP.
16 **"It is with great pleasure":** Theodore Roosevelt to Percival Lowell [n.d.], LPFP.
16 **"[Percival] is responsible":** Francis Peabody Jr. to Augustus Lowell, November 2, 1882, LPFP.
16 **monumental family quarrel:** Francis Peabody Jr. to Augustus Lowell, February 12, 1883; Katharine B. Lowell to Abbott Lawrence (brother), February 14 [1883]; and Augustus Lowell to Katharine B. Lowell, February 22, 1883, LPFP.
16 **"Oh mother forgive me":** Percival Lowell to Katharine B. Lowell, July 8, 1882, LPFP.
17 **Fifth Avenue Hotel:** *Harper's Weekly*, August 10, 1907, 1162–1164, 1175.
17 **"Women, children, and men":** *The Sun* (New York), September 19, 1883, 3. More at *The Evening Telegram* (New York), September 18, 1883, Third Edition, 1; *The Daily Graphic* (New York), September 18, 1883, Fifth Edition, 560; *The New-York Times*, September 19, 1883, 4 and 8; *New-York Tribune*, September 19, 1883, 8; *The New York Herald*, September 19, 1883, 5; *The Evening Post* (New York), September 19, 1883, Second Edition, 1.
17 **"like watching a kaleidoscope":** *The Sun* (New York), September 18, 1883, 1.

17 **Korea's first diplomatic mission:** Walter (1969).
18 **"ruddy-faced young American":** *The Sun* (New York), September 19, 1883, 3.
18 **"foreign secretary":** *The New-York Times*, September 18, 1883, 5; *The World* (New York), September 18, 1883, 5.
18 **"I am for the nonce a Corean":** Percival Lowell to Barrett Wendell, August 31 [1883], WFP.
18 **"There has until recently":** *New-York Tribune*, September 18, 1883, 8.
19 **"His parents knew nothing":** Catherine Elizabeth Peabody Gardner to John L. Gardner Jr., September 13 [1883], JLGP.
19 **"as imagination might paint":** P. Lowell (1886:43). For more on Lowell's visit to Korea, see Pai (2016).
19 **"so high in the heavens":** P. Lowell (1886:93).
19 **"For any people to write backwards":** P. Lowell (1886:107).
20 **"an alien race":** *The Atlantic Monthly*, May 1886, 696.
20 **sportsmanship at polo:** *The Boston Sunday Globe*, November 10, 1889, 7; *The Boston Herald*, September 17, 1890, 8; Stimson (1931:178).
20 **Gilded Age parties:** *The Boston Sunday Globe*, June 24, 1888, 13; *The Boston Daily Globe*, January 28, 1892, 5.
20 **best man for the marriage of Edith Wharton:** Benstock (1994:56).
20 **"His view of happiness":** Basil Hall Chamberlain to Lafcadio Hearn, January 10, 1893, in Koizumi (1936:1).
20 **"I would willingly have chloroformed":** P. Lowell (1891:37).
20 **"Sometimes a most disreputable Derby":** P. Lowell (1891:37–38).
21 **anthropologists placed societies on a gradient:** Powell (1888).
21 **"half civilized":** P. Lowell (1888:6).
21 **"There is no doubt":** *The New York Herald*, January 3, 1892, 7. See also C. Flammarion (1892a:110).

CHAPTER 2: THE FRENCH PHILOSOPHER

22 **"All are awaiting the birth":** Tuchman (1966:76).
22 **"'Fin de siècle'":** Wilde (1891:266).
23 **Anne-Émilie-Clara Goguet Guzman:** For details of her life, death, and bequest, see Winter (1984); Dumas (1888); and *The Westminster Review*, October 1892, 455–456.
23 **"An old lady has just died":** *The Examiner* (San Francisco), July 13, 1891, 3. Mme. Guzman's will stipulated that communication with one notable celestial body was ineligible for the prize: "I exclude the planet Mars which seems sufficiently known." *Transactions of the New York Academy of Sciences* 11 (1891): 4. This critical proviso, however, was largely overlooked by journalists, scientists, and the public.
23 **"The prize is to be named":** *The Chicago Daily Tribune*, September 14, 1891, 9.
23 **"The venerable lady of Pau":** *L'Astronomie*, November 1891, 403.
23 **"Astronomy plunges us into":** *The New York Herald*, June 28, 1891, 14.
24 **a thin volume:** The first edition numbered just 54 pages. By its second edition, the book ballooned to 570 pages. Crowe (1986:378).
24 **"Life is a law of nature":** C. Flammarion (1911:204).
24 **One of his novels, unabashedly autobiographical:** C. Flammarion (1891).
24 **gave Flammarion a château:** Cotardière and Fuentes (1994:162–164).
24 **jewels and artwork:** *The Evening Star* (Washington, D.C.), November 28, 1908, Part 3, 3.
24 **"She begged me to send you":** Blumenthal (1955:87). Also Blumenthal (1932:121). Both Camille and Sylvie Flammarion often recounted the strange tale of the gift of human skin, although in some versions of the story they clearly fictionalized

NOTES ON SOURCES

details. See Mme. C. Flammarion (1893); *The New York Herald*, January 16, 1893, 7; *La Chronique Médicale*, July 1, 1925, 205; and Thompson (1968:144–145).

26 **RELIURE EN PEAU HUMAINE:** Flammarion evidently bound several of his works in human skin, a not-unheard-of practice in the nineteenth century. The book I saw in the library at Juvisy, a copy of *La Pluralité des Mondes Habités*, was made from the skin of someone other than the young countess. Hers was used for a copy of Flammarion's *Les Terres du Ciel*. See Blumenthal (1932).

26 **"with the highest degree of probability":** Kepler (1965:42).

26 **"little black people, scorch'd":** Crowe (1986:19).

26 **"promise not to call me a Lunatic":** Crowe (1986:63).

26 **"the sun is richly stored":** Herschel (1795:104).

26 **"It is the opinion of all the modern philosophers":** Franklin (1961:345).

26 **almost 22 trillion:** Crowe (1986:199).

27 **"Well! This reasoning of savants":** *The New York Herald* (European Edition, Paris), June 12, 1892, 4. Also *The New York Herald*, May 29, 1892, 30.

28 **HOW TO TALK WITH THE FOLKS ON MARS:** *The New York Herald*, January 3, 1892, 7. See also C. Flammarion (1892a).

28 **"All this is bosh and nonsense":** *The Examiner* (San Francisco), July 13, 1891, 3.

29 **an omen or a deity:** Sheehan and O'Meara (2001:47–48); Proctor (1881:278).

30 **"Nothing can exceed the calm, fierce, golden":** Whitman (1892:415).

31 **"[Mars] offers so many points of resemblance":** *The New York Herald*, January 3, 1892, 7. See also C. Flammarion (1892a:108).

31 **"In the 400th year from Columbus":** William H. Pickering to his mother, July 31, 1892, WHPP, Box 1.

32 **"superior to and more powerful":** Wright (1987:28).

32 **"the Mars boom of 1892":** *The Observatory*, September 1894, 312. See also *The Edinburgh Review*, October 1896, 368.

32 **"I wanted to put into each issue":** Barrett (1941:81).

33 **COME, VISIT MARS!:** *The World* (New York), August 2, 1892, 1.

33 **MARS AND ITS MEN:** *The World* (New York), August 5, 1892, 1.

33 **REMARKABLE REVELATIONS MADE:** *The World* (New York), August 2, 1892, 1.

33 **"Steadily and brilliantly these three effulgent lights":** *The World* (New York), August 4, 1892, 2.

CHAPTER 3: WHAT THE COLORBLIND ASTRONOMER SAW

34 **"Three children remain":** *The New-York Times*, July 4, 1891, 8. For more on Alfred Roosevelt's accident, see *The Evening Post* (New York), July 2, 1891, Last Edition, 12; *The New-York Times*, July 3, 1891, 1; *The World* (New York), July 3, 1891, 1; *The Sun* (New York), July 3, 1891, 2; *The Evening Post* (New York), July 3, 1891, Last Edition, 1; and *The Port Chester Journal*, July 9, 1891, 4.

35 **"of the old Lowell family":** *The New York Herald*, July 4, 1891, 9.

35 **studying Japanese "occult" practices:** P. Lowell (1895c).

35 **"I shall out of regard for the family's wishes":** Percival Lowell to William Lowell Putnam, September 24, 1891, LELP.

35 **resumed polo:** *The Boston Sunday Globe*, April 17, 1892, 21; *The Boston Times*, April 24, 1892, 8.

35 **attended cultural exhibitions:** *The Boston Daily Globe*, January 28, 1892, 5; *Boston Daily Advertiser*, January 28, 1892, 1.

35 **graced dinner dances:** *The Boston Sunday Globe*, February 7, 1892, 21.

35 **"the event of the season":** *The Boston Daily Globe*, January 9, 1892, 4; *The Boston Sunday Globe*, January 10, 1892, 21.

35 **meetings of the Boston Society of Natural History:** *The Boston Daily Globe*, January 7, 1892, 4; *The Boston Daily Globe*, April 7, 1892, 2.

NOTES ON SOURCES

35 **"I have thought of some things"**: Percival Lowell to William Lowell Putnam, July 22, 1891, LELP.
35 **"The miseries of the dog-days"**: *Boston Evening Transcript*, August 5, 1892, 4.
35 **IS THE PLANET INHABITED**: *The Herald Advance* (Milbank, SD), August 19, 1892, 7.
35 **CONCLUSIVE EVIDENCE THAT THE PLANET**: *The Wheeling Daily Intelligencer*, August 4, 1892, 6.
36 **"It is not impossible"**: *The Press* (New York), August 4, 1892, 4.
36 **sun glinting off snowcapped peaks or off clouds**: W. W. Campbell (1894); Holden (1894a).
36 **Even Camille Flammarion**: C. Flammarion (1897:316).
36 **"The latest astronomical news"**: *The Theatre*, Summer 1892, 201–202.
37 **a consequential telescope**: Carpino (2017).
38 **"I completely forgot I was married!"**: E. Schiaparelli (1954:25).
38 **"I have adopted the rule"**: *Proceedings of the American Association for the Advancement of Science for the Forty-Third Meeting Held at Brooklyn, N.Y. August 1894*, 30.
38 **Schiaparelli turned his telescope**: Sheehan and Bell (2021:120).
38 **He was colorblind**: Sheehan (1997).
39 **as if conducting a land survey**: C. Flammarion (2015:245–246).
39 **"They may disappear wholly"**: Van Biesbroeck (1934:556–557).
39 **a phenomenon he called *gemination***: G. Schiaparelli (1894:720–721).
39 **"What could all this mean?"**: Terby (1888:285).
40 **CANALS ON THE PLANET MARS**: *The Times* (London), April 13, 1882, 12.
40 **"Schiaparellophobomania"**: François Terby to Giovanni Schiaparelli, September 17, 1892, in G. Schiaparelli (1976:97).
40 **"It is perhaps hardly yet safe"**: *The Forum*, September 1890, 86–87.
40 **Flammarion's observatory in France**: *The New York Herald*, August 14, 1892, 17.
40 **Harvard's in Peru**: Pickering (1892).
40 **the Lick in California**: Holden (1892).
40 **"What is going on in Mars?"**: *The New York Herald* (European Edition, Paris), August 13, 1892, 2.
40 **"a gigantic semaphore"**: *The World* (New York), August 2, 1892, 1.
41 **"Who shall say that some day"**: *The Boston Daily Globe*, August 4, 1892, 4.
41 **"Marsites"**: *The Pittsburgh Post*, August 4, 1892, 2.
41 **"Martials"**: *The Spectator* (London), August 13, 1892, 218.
41 **"Pshaw!"**: *The Morning Call* (San Francisco), August 4, 1892, 1.
41 **"the Czar" and "the Dictator"**: Osterbrock (1984).
41 **"All, or nearly all"**: *The Morning Call* (San Francisco), August 3, 1892, 1.
41 **"The prevalent excitement"**: *The New York Herald*, August 3, 1892, 3.
41 **"I am by no means ready"**: *The Examiner* (San Francisco), August 2, 1892, 1.
42 **"'I do not know!'" screamed an exasperated Flammarion**: *The New York Herald*, October 6, 1890, 7. Flammarion made these comments in 1890, in response to published remarks by Holden that were almost identical to those of 1892.
42 **"There are [conservatives] even in science"**: Tyndall (1870:6).
42 **"Imagination is the soul of science"**: *The Japan Weekly Mail*, April 6, 1889, 341.
43 **"imagined themselves for a time"**: *The Boston Herald*, October 19, 1892, 6.
43 **created the illusion of spaceflight**: *The Electrical Engineer*, March 9, 1892, 254–255; *Scientific American*, April 9, 1892, 229.
43 **"Slowly as time goes by"**: *The Examiner* (San Francisco), October 1, 1893, 13.
43 **"a gentleman long and favorably known for his popular articles"**: *Boston Evening Transcript*, October 19, 1892, 5.
43 **"Looking at such a landscape"**: Serviss (1892b:12).
44 **"In his use of imagination Mr. Serviss"**: *The Brooklyn Daily Eagle*, May 16, 1895, 10.

NOTES ON SOURCES   259

44   "The results of the observations": Serviss (1892a).
44   "no reputable astronomer would give": *The Sun* (New York), August 14, 1892, Second Section, 7.
44   "But nothing has yet been proposed": Serviss (1889:55).
45   "That there is something very singular": *The Sun* (New York), August 14, 1892, Second Section, 7.
45   He remembered gazing at Mars: P. Lowell (1906:32); A. L. Lowell (1935:5); *The Sunday Herald* (Boston), May 17, 1914, Special Features Section, 6; Percival Lowell to Hector Macpherson, November 16, 1903, LOAC, Copybook 5E, 410.
45   "Could you kindly tell me": Percival Lowell to Edward C. Pickering, September 9, 1892, RHCOD, Box 169, Folder 7.
45   "Last night I observed Saturn": Percival Lowell to William Lowell Putnam [n.d. but circa February 1, 1893], LELP.
45   declaring: "At once!": Leonard (1921:41).
45   It said, simply, "Hurry": Strauss (2001:178).

CHAPTER 4: MARS HILL

46   "mythical and poetic geography": Giovanni Schiaparelli to Otto Struve, January 4, 1878, in G. Schiaparelli (1963:11).
46   "like the wail of a lost soul": P. Lowell (1906:8).
46   "Seeing growing dim": Entry for June 7, 1894, HLLO, Logbook 1, 26.
47   "and, after their travel": P. Lowell (1895a:86–87).
47   "With the best will in the world": Entry for June 19, 1894, HLLO, Logbook 1, 71. Also in P. Lowell (1898:186).
48   "I find it very difficult to raise the money": William H. Pickering to John Brashear, November 1, 1893, RHCOD, Box 35, Letterbook E4.
48   "Mills, factories, furnaces, mines": Rezneck (1953:325).
48   At a dinner party in Boston: "W. H. Pickering . . . was invited to dine with Percival Lowell at the house of a friend, Mr. Foncelli. At this dinner, Pickering described our successful South American work on Mars and Lowell became very much interested indeed. He proposed to finance an expedition that Pickering thought should go to Arizona." A. E. Douglass, "Lowell Observatory (Reminiscences)," 1–2, AEDP, Box 17. For further background on the early collaboration among Lowell, Pickering, and Douglass, see B. Z. Jones and L. G. Boyd (1971:325–330); Strauss (1994); and Strauss (2001:173–186).
48   "Mr. Percival Lowell . . . now comes to the rescue": *The Boston Herald*, February 13, 1894, 4.
48   two coffinlike boxes: A. E. Douglass, "Lowell Observatory (Reminiscences)," 3, AEDP, Box 17.
48   Lowell's "far travelled telescope": *Boston Commonwealth*, May 26, 1894, 3; *Nature*, June 14, 1894, 149.
49   "Tombstone will not be suitable for ladies": A. E. Douglass to Percival Lowell, March 9, 1894, PLP, Percival Lowell Correspondence.
49   "centipedes, scorpions, tarantulas, rattlesnakes": A. E. Douglass to Percival Lowell, April 12, 1894, PLP, Percival Lowell Correspondence.
49   "those who remain sleep out doors": A. E. Douglass to Percival Lowell, March 16, 1894, PLP, Percival Lowell Correspondence.
49   "Flagstaff is a town of 1200": A. E. Douglass to Percival Lowell, April 7, 1894, PLP, Percival Lowell Correspondence.
49   the nickname "Skylight City": Cline (1994:59).
50   "fair stores and many pleasant people": A. E. Douglass to Percival Lowell, April 12, 1894, PLP, Percival Lowell Correspondence.

NOTES ON SOURCES

50 **"Barbarism all along the line":** Cline (1976:132).
50 **he had decided: "Flagstaff it is":** Percival Lowell to A. E. Douglass, April 16, 1894, PLP, Percival Lowell Correspondence.
50 **"In answer to name":** Percival Lowell to A. E. Douglass, March 15, 1895, PLP, Percival Lowell Correspondence.
50 **"a gentleman long and favorably known as a Japanese scholar":** Clipping from an unidentified Boston newspaper, circa March 1894, PLP, Lowell Scrapbook 1894–1897.
51 **"With regard to the observatory's plan of work":** *Boston Commonwealth*, May 26, 1894, 3. The text printed in the paper came directly from the manuscript of Lowell's talk, as noted in *Boston Commonwealth*, July 7, 1894, 3.
51 **Lowell displayed to his audience a globe:** *Boston Commonwealth*, May 26, 1894, 4.
51 **"The foregoing words seem to me":** Holden (1894b:163).
52 **"Holden seems like a fly":** Katharine B. Lowell to Percival Lowell, August 20 [1894], PLP, Percival Lowell Correspondence.
52 **"Good-bye my darling":** Katharine B. Lowell to Percival Lowell [n.d. but circa August 20, 1894], PLP, Percival Lowell Correspondence.
52 **"the young Flagstaff band":** A. L. Lowell (1935:75).
52 **divided the night into watches:** *The Daily Inter Ocean* (Chicago), March 9, 1895, 10.
52 **"Are you observatory people?":** A. L. Lowell (1935:74).
52 **"Snow-cap exceedingly small":** Entry for August 22, 1894, HLLO, Logbook 1, 204. Also in P. Lowell (1898:17).
53 **"No snow; certainly to speak of":** Entry for October 14, 1894, HLLO, Logbook 2, 38. Also in P. Lowell (1898:19).
53 **"So much water suddenly produced":** P. Lowell (1894:550).
53 **"Suspicions of canals":** Entry for June 13, 1894, HLLO, Logbook 1, 52.
53 **"To look for the canals":** P. Lowell (1895a:140).
53 **"These sudden revelation peeps":** Entry for June 9, 1894, HLLO, Logbook 1, 35. Also in P. Lowell (1898:186).
53 **"Not sufficient time to make sure":** Entry for June 20, 1894, HLLO, Logbook 1, 75.
53 **"The number of canals increases":** A. L. Lowell (1935:73).
54 **seeing the lines "in profusion":** Entry for November 2, 1894, HLLO, Logbook 2, 118. Also in P. Lowell (1898:85).
54 **"It looks as if it had been cobwebbed":** P. Lowell (1906:175).
54 **smoke from Flagstaff's chimneys rose straight upward:** P. Lowell (1898:206).
54 **pointed his telescope at Mars and nearly gasped:** P. Lowell (1906:193).
54 **"The canals of Mars have begun to double":** *Boston Evening Transcript*, November 19, 1894, 1; *The New-York Times*, November 20, 1894, 16.
54 **"Now you see that the more one observes Mars":** *New-York Tribune*, December 1, 1894, 2. See also *The English Mechanic*, November 16, 1894, 298.
54 **five inches from his face:** *New-York Tribune*, December 12, 1895, 6. For more on Serviss's visit with Schiaparelli, see *New-York Tribune*, August 31, 1894, 4, and *The Standard Union* (Brooklyn), October 3, 1894, 2.
54 **gradually going blind:** Giovanni Schiaparelli to Percival Lowell, March 27, 1903, PLP, Percival Lowell Correspondence. Transcribed and translated in J. Putnam and W. Sheehan (2021:196).
55 **"Our own little planet is proving itself":** *Democrat and Chronicle* (Rochester, NY), September 5, 1894, 6.
55 **"Nothing, in truth, so much exalts":** Arnold (1894:411).
55 **"We have enlarged enormously":** Arnold (1894:413).
55 **"The best thing that could happen":** Arnold (1894:410).
56 **an ode to the shipwrecked vessel:** A. L. Lowell (1935:4).
56 **In Japan, he wrote poems:** *Scribner's Magazine*, September 1888, 365; *The Boston Post*, June 28, 1889, 1–2.

NOTES ON SOURCES

56 **friends advised that he stick to prose:** Strauss (2001:78).
56 **"A sister planet, whose sister face":** Percival Lowell, "Mars," PLP, Unpublished Lectures, Manuscripts, and Poems, Box 1U, Folder 5. There are two slightly different versions of the poem. I have quoted from what appears to be the later, more polished one.
57 **"Are these very survivors dead?":** Percival Lowell, "Mars," PLP, Unpublished Lectures, Manuscripts, and Poems, Box 1U, Folder 5.

CHAPTER 5: LOST WORLD

58 **"palace for the people":** Boston Public Library (1889:6). For news of the library's opening, see *The Boston Post*, February 2, 1895, 1, and February 4, 1895, 2; *The Daily Inter Ocean* (Chicago), February 9, 1895, 10; and *The Times-Democrat* (New Orleans), February 10, 1895, 15.
58 ***The Quest and Achievement of the Holy Grail:*** *The Sunday Herald* (Boston), May 12, 1895, 31.
59 **Ticketholders quickly found their seats:** "The Lowell Institute is the only place at which people are always and invariably punctual." Lilian Whiting, in *The Daily Inter Ocean* (Chicago), February 22, 1896, 16.
59 **filling the rows:** "Mr. Lowell's four lectures on the planet Mars were heard by crowded audiences of people who filled every seat and all the standing-room in Huntington Hall." *Boston Commonwealth*, February 9, 1895, 1; reprinted in *Scientific American*, March 2, 1895, 137.
59 **Promptly at seven forty-five:** "When the lecturer comes forward the entire house is seated, composed, quiet, and in its right mind. . . . The lectures are all precisely an hour in length—beginning at a quarter before 8 o'clock and ending promptly at a quarter before 9." Lilian Whiting, in *The Daily Inter Ocean* (Chicago), February 22, 1896, 16.
59 **"Amid the seemingly countless stars":** *The Atlantic Monthly*, May 1895, 594.
60 **"Now, to all forms of life":** *The Atlantic Monthly*, May 1895, 596.
60 **"That beings constituted physically":** *The Atlantic Monthly*, May 1895, 603.
60 **"The lecture proved very interesting":** *The Boston Herald*, January 24, 1895, 8.
60 **"After air, water":** *The Atlantic Monthly*, June 1895, 749.
61 **"Several important facts conspire":** *The Atlantic Monthly*, June 1895, 751.
61 **"[Mars] is relatively more advanced":** *The Atlantic Monthly*, June 1895, 756.
61 **"Now, if a planet were at any stage":** *The Atlantic Monthly*, June 1895, 758.
62 **"At this point in our inquiry":** *The Atlantic Monthly*, June 1895, 758.
62 **Boston *Sunday Herald* offered a summary:** *The Sunday Herald* (Boston), January 27, 1895, 5.
62 **one of Percival Lowell's favorite authors:** Basil Hall Chamberlain to Lafcadio Hearn, August 5, 1893, in Koizumi (1936:33), and June 23, 1893, in Koizumi (1937:82).
62 **"Where will the romance writers":** *The African Review*, June 9, 1894, 762.
63 **lantern slides that showed stylized drawings:** *The Boston Herald*, January 31, 1895, 6.
63 **"At times the canals are invisible":** *The Atlantic Monthly*, July 1895, 113.
63 **"Not until such melting has progressed":** *The Atlantic Monthly*, July 1895, 115.
63 **fifteen miles across at a minimum:** *The San Francisco Call*, September 6, 1896, 9. See also Serviss (1896).
63 **Lowell had himself seen this phenomenon:** Percival Lowell to Elizabeth Lowell, April 8 [1877], LELP.
63 **"A solution of their character":** *The Atlantic Monthly*, August 1895, 227.
63 **"A mind of no mean order":** *The Atlantic Monthly*, August 1895, 234.
64 **"All this, of course":** *The Atlantic Monthly*, August 1895, 231.

64 **"It is impossible in print"**: *Boston Commonwealth*, February 9, 1895, 1; reprinted in *Scientific American*, March 2, 1895, 137.
64 **"It is as fascinating as a romance"**: *The Daily Inter Ocean* (Chicago), June 8, 1895, 16.
64 **Lilian Whiting Club:** *The Times-Democrat* (New Orleans), October 17, 1897, 18.
65 **"There's science and psychology"**: *The Courier-Journal* (Louisville), November 27, 1904, Section 2, 4. See also *The Washington Post*, November 27, 1904, Fourth Part, 2.
65 **among those at the Tremont Theatre:** *The Times-Democrat* (New Orleans), October 23, 1892, 15.
65 **"What data have you deduced"**: *The Daily Inter Ocean* (Chicago), March 9, 1895, 10. See also *The Times-Democrat* (New Orleans), March 10, 1895, 17.
65 **"What could more impressively lead"**: Clipping headlined THE WORLD BEAUTIFUL, source unidentified, 1895, PLP, Lowell Scrapbook 1894–1897.
65 **"My lectures are coming out"**: Percival Lowell to A. E. Douglass, March 23, 1895, PLP, Percival Lowell Correspondence.
65 **"the entire first edition"**: *The Daily Inter Ocean* (Chicago), May 11, 1895, 10.
66 **"The rather wild speculations of Flammarion"**: Clipping headlined IS MARS INHABITED?, source unidentified, circa August 1895, PLP, Lowell Scrapbook 1894–1897.
66 **commencement addresses that June:** *Poughkeepsie Daily Eagle*, June 13, 1895, 6; *The San Diego Sun*, June 22, 1895, 5; *Paterson Evening News* (Paterson, NJ), June 26, 1895, 1.
66 **"Is there anything to prevent"**: *Washington County Post* (Cambridge, NY), June 28, 1895, 4.
66 **"an army of 200 millions of men"**: *The Journal of the British Astronomical Association* 5 (1894–1895): 209.
66 **"It is a point to be noted"**: *Publications of the Astronomical Society of the Pacific*, April 1, 1895, 118.
66 **"cheapen the science of astronomy"**: *The Sun* (New York), December 10, 1895, 6.
66 **"the good Sir Percival"**: George Ellery Hale to Wright, October 16, 1901, in Strauss (2001:313). Note: I was unable to locate the original letter in the archival collection cited.
66 **"There is no danger that astronomy"**: *The Sun* (New York), December 10, 1895, 6.
66 **"Though this article"**: *The Sunday Oregonian* (Portland), July 28, 1895, 4.
67 **"[His] theory accounts very well"**: Serviss (1896).
67 **"men or creatures possessing the attributes of men"**: Clipping headlined MARS AND THE MARTIANS, source unidentified, circa August 1895, PLP, Lowell Scrapbook 1894–1897.
67 **"What a colossal picture"**: *Cosmopolis*, July 1896, 79.
67 **"At the close of this nineteenth century"**: *The Edinburgh Review*, October 1896, 384. Although unsigned, the article is widely attributed to Clerke. See Wallace (1907:21).
68 **on stage at the Tremont:** *The Boston Daily Globe*, October 2, 1895, 5.
68 **"sumptuously bound in scarlet and gold"**: *The Daily Inter Ocean* (Chicago), January 25, 1896, 16.
68 **"On the rostrum or the page"**: *The Brooklyn Citizen*, January 26, 1896, 15.
68 **"I have studied [them]"**: *Harper's New Monthly Magazine*, September 1896, 636.
68 **Lowell's book soon found its way into the collection:** *Monthly Bulletin of Books Added to the Public Library of the City of Boston*, January–April 1896, 20.
68 **"He is a scholar by instinct"**: *The Boston Sunday Globe*, December 29, 1895, 24.
68 **"Fault of the pilot"**: A. L. Lowell (1935:92). For more on the grounding of the steamship *Spree*, see *The Daily News* (London), December 20, 1895, 7.

NOTES ON SOURCES   263

CHAPTER 6: ALLIES AND ADVERSARIES

69 **the décor left no doubt:** For descriptions of Flammarion's apartment, see Sherard (1894); Griffith (1896); A. L. Lowell (1935:93); and *The Morning Call* (San Francisco), May 28, 1893, 35–36.
70 **"There is where I put his hair":** Fuller (1913:112).
70 **"Sir, it's an American astronomer":** All quotations and other details from Flammarion's first meeting with Lowell come from *L'Illustration*, January 18, 1896, 59–60. Portions of the article were later published, in English, in *The New-York Times*, February 9, 1896, 29, and *Scientific American*, February 29, 1896, 133–134.
70 **The two men had corresponded:** Percival Lowell to Camille Flammarion, July 19, 1894; June 11 and July 25, 1895, CFP.
72 **the *canalists* and the *anti-canalists*:** These labels were not in widespread use at the time but were applied in hindsight to the competing camps. Among the first to employ the terms in print was Lepper (1919:20).
73 **"I will risk all the horrible suspicions":** *The Observatory*, April 1896, 177. Although this column ("From an Oxford Note-Book") was written anonymously, the author was H. H. Turner. See *Popular Astronomy*, February 1931, 61.
73 **"I suspected them myself":** Quoted in Percival Lowell to Simon Newcomb, March 15, 1903, SNP, Box 30. See also P. Lowell (1903:60) and P. Lowell (1906:249).
73 **"Here, the intervention of intelligent thought":** G. Schiaparelli (1898:428).
73 **"*Semel in anno licet insanire*":** G. Schiaparelli (1898:429).
74 **"Palms choke the outlook":** Percival Lowell to William Lowell Putnam and Elizabeth Lowell Putnam, January 13, 1896, LELP.
74 **"On arriving at one of these 'oases'":** *The World* (New York), January 17, 1897, Sunday Magazine, 24.
75 **illustration portraying the Martians as if ancient Romans:** *The Sunday Post* (Boston), May 31, 1896, 24.
75 **"Mars ought to appreciate the devotion":** *The Two Republics* (Mexico City), February 14, 1897, U.S. Edition, 5.
75 **"I have no doubt that there is life":** *The Chicago Times-Herald*, July 20, 1896, 7.
75 **train robbery, a murder, even the dynamiting:** *The Coconino Weekly Sun* (Flagstaff), July 23, 1896, 1 and 2.
75 **"The significant point of the whole":** *The New York Herald*, October 11, 1896, Sixth Section, 7.
75 **may have lived with bipolar disorder:** Sheehan (2024:314); Sheehan (2019:372).
76 **"This looks to me suspiciously like Mars":** *The Observatory*, December 1896, 420.
76 **"I have no hesitation in saying":** Holden (1897).
76 **"We cannot help considering":** Antoniadi (1898a:316). See additional implied criticism of Lowell's Venus observations in Antoniadi (1897:45–46). E. M. Antoniadi would later play a central role in discrediting Lowell's Mars work. See chapter 17.
76 **blood vessels of his own retina:** Sheehan and Dobbins (2003).
77 **eventually acknowledge his error:** P. Lowell (1902).
77 **"I have had what I never supposed":** Percival Lowell to A. E. Douglass, April 21, 1897, PLP, Percival Lowell Correspondence.
77 **"Am pretty weak," he confessed:** Percival Lowell to A. E. Douglass, September 20, 1897, PLP, Percival Lowell Correspondence.
77 **condition widely known then as *neurasthenia*:** Oppenheim (1991).
77 **forced him into bed for a month:** A. L. Lowell (1935:98).
77 **Lounging at the Hamilton Hotel:** *Home Journal* (Boston), January 1, 1898, 4.
77 **"*Festina lente*":** A. L. Lowell (1935:99).

77 **"Everyone truly hopes"**: Giovanni Schiaparelli to Percival Lowell, December 15, 1898, GSP. Transcription and alternate translation in J. Putnam and W. Sheehan (2021:190–191).
78 **"rid the Observatory of its incubus"**: William J. Hussey to E. L. Campbell, July 7, 1897, HFP, Box 3, "Holden (E.S.) and University of California, 1897."
78 DIRECTOR HOLDEN RESIGNS: *The San Francisco Call*, September 29, 1897, 12.

CHAPTER 7: WAR STORIES

79 **"No phase of anthropomorphism"**: *The Saturday Review*, April 4, 1896, 346. Regarding Wells's authorship, see Hughes and Philmus (1973:112).
80 **"Concerning the Nose"**: *The Ludgate*, April 1896, 678–681.
80 **"The Living Things That May Be"**: *The Pall Mall Gazette*, June 12, 1894, 4. Regarding Wells's authorship, see Hughes and Philmus (1973:106).
80 **"Through a Microscope"**: *The Pall Mall Gazette*, December 31, 1894, 3. Also in Wells (1898a:238–245).
81 **"Suppose some beings from another planet"**: *The Strand Magazine*, February 1920, 154.
81 *What planet should I choose?*: *The Daily News* (London), January 26, 1898, 6.
81 **"They are on every street"**: *The Daily Inter Ocean* (Chicago), June 15, 1895, 16.
81 **"spinning down" a boulevard**: *The New York Herald* (European Edition, Paris), April 11, 1897, 4.
81 CANALS ARE REALLY BICYCLE PATHS: *The Morning Times* (Washington, D.C.), June 7, 1896, 17.
81 **"I'm doing the dearest little serial"**: H. G. Wells to Elizabeth Healey, late spring 1896, in Wells (1998:261).
82 **"I don't know if you see *Pearson's Magazine*"**: H. G. Wells to Fred Wells, December 31, 1896, in Wells (1998:281).
82 **"No one would have believed"**: *The Cosmopolitan*, April 1897, 615. Minor changes to the wording and punctuation were made when the story was published in book form. See Wells (1898b:3–4).
83 **"There is no more absorbing serial"**: *The Globe* (Toronto), October 30, 1897, 6.
83 **"The present installment is rather grewsome"**: *The Evening Times* (Washington, D.C.), April 3, 1897, 4.
83 **It was a tub of butter**: *The Duluth News Tribune*, November 11, 1897, 5.
84 **Martians *are* what we humans *will be***: *The Daily News* (London), January 26, 1898, 6.
84 **"The Man of the Year Million"**: *The Pall Mall Gazette*, November 6, 1893, 3. Republished under the title "Of a Book Unwritten" in Wells (1898a:161–171).
85 **"Had our instruments only permitted it"**: *The Cosmopolitan*, April 1897, 616.
85 **"In strict confidence," Roosevelt wrote**: Theodore Roosevelt to Col. F. V. Greene, September 23, 1897, FVGP.
85 **cunning young publisher William Randolph Hearst**: For background on Hearst and his entry into the New York newspaper business, see Tebbel (1952); Churchill (1958); and Nasaw (2000).
86 **Arthur Brisbane, a shrewd editor**: O. Carlson (1937).
86 **participated in *The World*'s famously hyped Mars coverage**: *The World* (New York), July 31, 1892, 6.
86 **day after day for a month**: The story ran in the *New York Evening Journal* from December 15, 1897, to January 11, 1898.
86 **without the knowledge or permission of H. G. Wells**: *The Saturday Review*, February 12, 1898, 217; *The Critic*, March 12, 1898, 184.
87 **"collapsed itself and buried beneath its ruins"**: *The Boston Post*, January 22, 1898, 5.
87 **"Let us go to Mars"**: *New York Evening Journal*, January 12, 1898, 9.

87 **"What's all this talk about Uncle Sam"**: *New York Evening Journal*, February 26, 1898, 7.
88 **"McKinley's War of the Worlds"**: *The Age-Herald* (Birmingham, AL), May 29, 1898, 9; *The Sunday Post* (Boston), May 29, 1898, War Extra, 13.
88 **"Oh, but we have had a bully fight"**: *The New York Herald*, August 16, 1898, 5.
88 **"As we see to-day, in spite"**: *The Evening Post* (New York), March 14, 1898, Last Edition, 6.
88 **"Do you think the canals upon Mars"**: *The Sunday Chronicle* (Chicago), September 25, 1898, 38.

CHAPTER 8: WIRELESS

90 **Mark Twain and other celebrities**: R. U. Johnson (1923:400–401).
90 **veritable sorcerer's den**: Stephenson (1895); Tesla (1899); *The New-York Times*, March 31, 1895, 13.
90 **ribs, shoulders, heart, skull**: Tesla (1896).
91 **engaged in self-experimentation**: *The Detroit Free Press*, February 16, 1896, 16; *Electrical Review*, May 5, 1897, 207; *New York Journal and Advertiser*, August 4, 1897, 1; *New York Journal and Advertiser*, October 9, 1898, 20.
91 **"You see," he explained**: *The Evening Star* (Washington, D.C.), June 6, 1896, 15.
91 **"Ain't science wonderful!"**: Milton (1989:44).
91 **early booster of Tesla's**: *The World* (New York), July 22, 1894, 17.
91 **the more bizarre, the better**: Stevens (1991:83–84); W. J. Campbell (2001:12).
91 **TESLA SAYS MEN MAY YET LIVE**: *The World* (New York), September 5, 1897, 25.
91 **TESLA'S ELECTRICAL SUBSTITUTE**: *New York Journal and Advertiser*, October 9, 1898, 20.
91 **TESLA'S LATEST DISCOVERY**: *The World* (New York), January 3, 1897, 17.
91 **TESLA'S GREAT INVENTION**: *New York Journal and Advertiser*, May 1, 1898, 24–25.
91 **"Whole pages of the yellow journal"**: *Scientific American*, November 26, 1898, 338.
91 **no time for romantic involvement with women**: *The Evening Star* (Washington, D.C.), June 6, 1896, 15.
92 **his affinity for men**: W. B. Carlson (2013:239–240); Seifer (1998:412).
92 **Marconi flung his electrical signals**: Moffett (1899).
92 **the massive Auditorium Building**: Garczynski (1891); Kirkland and Kirkland (1894:362–364); Siry (2002).
92 **Chicago's capitalist elite**: This was a meeting of the city's Commercial Club. See *Modern Machinery*, March 1, 1900, 81.
93 **"the magician of the electrical world"**: *The Chicago Sunday Tribune*, May 14, 1899, 1.
93 **"Only the addition of a pointed beard"**: *The Chicago Times-Herald*, May 15, 1899, 2.
93 **beef tenderloin and roast squab**: The banquet menu can be found in the Commercial Club's program for the evening of May 13, 1899, NTM, Personal Fund, "MNT, LXXXV, 194 A."
93 **"The power is stored within the boat"**: *St. Louis Post-Dispatch*, May 18, 1899, 15. For more on Tesla's demonstration before the Commercial Club, see *The Kansas City Times*, May 15, 1899, 5; *The Chicago Daily News*, May 15, 1899, Last Edition, 1; *Electrical Review*, May 17, 1899, 309; and *Western Electrician*, May 20, 1899, 285. According to some biographers, Tesla unveiled his radio-controlled boat a year earlier (in May of 1898) at an electrical exhibition at New York's Madison Square Garden, but this story is almost certainly myth, as noted by W. B. Carlson (2013:230, 451). Tesla's first display of the boat appears to have been in November 1898—at his New York laboratory—an event that received widespread press attention. See *The Sun* (New York), November 8, 1898, 7, and *The New York*

Herald, November 8, 1898, 6. Indeed, one report at the time (*New-York Tribune*, November 9, 1898, 8) referenced the earlier electrical show at Madison Square Garden but only to say that a rival inventor—*not Tesla*—had demonstrated a radio-activated *submarine mine* there, in contrast to Tesla's wirelessly controlled torpedo boat. See more in *The New York Times*, May 7, 1898, 12.

94 **"The invention of the torpedo boat"**: *The Chicago Times-Herald*, May 15, 1899, 1.

94 **MAYBE WE CAN TALK TO MARS**: *New York Journal and Advertiser*, May 18, 1899, 6.

95 **"Let him send three dots or dashes"**: *New York Journal and Advertiser*, May 18, 1899, 6. Tesla had proposed such an idea two years earlier; see *New York Journal and Advertiser*, June 13, 1897, 17.

95 **"Are you going to Colorado"**: *The Chicago Times-Herald*, May 15, 1899, 2.

95 **"light, dry, electrical atmosphere"**: Atchison, Topeka, and Santa Fe Railway Company (1898:10–14).

95 **"not a good country for hair"**: Tesla and Marinčić (1978:128).

95 **checked in at the Alta Vista Hotel**: Hunt and Draper (1964:106); *The Denver Sunday Post*, September 17, 1899, 24; *The Great West*, July 1896, 43.

95 **numbers divisible by three**: Tesla (1919a:746).

95 **"If he wants them, let him have them"**: Hunt and Draper (1964:107).

96 **GREAT DANGER – KEEP OUT**: Photograph in Tesla and Marinčić (1978:427). For descriptions of the laboratory, see *The New York Herald*, November 12, 1899, Fourth Section, 8, and *The Evening Telegraph* (Colorado Springs), June 2, 1899, 5.

96 **"filled with dynamos, electric wires"**: *The Evening Telegraph* (Colorado Springs), June 21, 1899, 5.

96 **"We all thought he was crazy"**: Hunt and Draper (1964:114).

96 **"Of course you cannot expect me"**: *The Rocky Mountain News* (Denver), May 17, 1899, 1.

96 **"the pulse of the globe"**: Tesla (1901).

96 *Click-click-click*: It is not entirely clear what the signal sounded like that night in Tesla's Colorado Springs laboratory. The inventor sometimes arranged his equipment so that an electromagnetic signal caused a telegraph sounder to click; at other times, it caused a tone to beep in a telephone receiver. W. B. Carlson (2013:274–275) presumes that Tesla used the latter setup on the night in question. However, Tesla's friend Julian Hawthorne evocatively described what Tesla heard as "three fairy taps . . . three soft impulses" (*The North American* [Philadelphia], January 11, 1901, 8), and Tesla said that his instrument "recorded certain feeble movements" (*The Evening Telegram* [New York], January 3, 1901, 10 O'Clock P.M. Edition, 7), suggesting that when the mysterious signals arrived, Tesla's telegraph sounder began tapping out gentle clicks in triplets.

97 **"We have had a wonderful industrial development"**: *The Chicago Daily Tribune*, October 11, 1899, 1–2.

97 **lunches with friends from Boston**: Among those with whom Lowell socialized was the Harvard psychologist William James. See A. L. Lowell (1935:100–101); Percival Lowell to David Todd, March 27, 1900, DPTP, Box 6, Folder 195; and William James to Theodora Sedgwick, March 21, 1900, in James (2001:173–175).

97 **another Massachusetts stargazer, David Todd**: Lowell and Todd had corresponded as early as 1895. See Percival Lowell to David Todd, August 5, 1895, DPTP, Box 6, Folder 195.

97 **"The explanation of the canals"**: D. Todd (1897:360).

97 **"Todd . . . is not a fossil"**: Percival Lowell to A. E. Douglass, August 9, 1897, PLP, Percival Lowell Correspondence.

97 **transcribed, assembled, and published the poems**: Dobrow (2018).

98 **passionate affair with Emily Dickinson's brother**: Longsworth (1984).

98 **"I cannot explain it"**: Entry for January 20, 1882, journal of Mabel Loomis Todd, 142, MLTP, Diaries and Journals Microfilm, Reel 8.

NOTES ON SOURCES   267

98 "Mr. Percival Lowell . . . is very charming": Entry for May 8, 1897, diary of Mabel Loomis Todd, MLTP, Diaries and Journals Microfilm, Reel 3.
98 *Total Eclipses of the Sun*: M. L. Todd (1894).
98 "Your book is on my table": Percival Lowell to Mabel Loomis Todd, [n.d., likely May 16, 1897, despite notation that says "ca. 1895"], MLTP, Box 17, Folder 445.
98 "We spend our lives chasing eclipses": *The Sunday Inter Ocean* (Chicago), September 2, 1900, 35.
98 The Todds cultivated wealthy benefactors: Dobrow (2017); M. L. Todd (1898).
99 "Mr. Lowell quite jolly": Entry for May 24, 1900, diary of Mabel Loomis Todd, MLTP, Diaries and Journals Microfilm, Reel 4.
99 "She is here as ubiquitous": Percival Lowell to Elizabeth Lowell Putnam, May 25, 1900, LFP, Box 3, Folder 42.
99 "the very spirit of interplanetary space": M. L. Todd (1912:126). For more on the total eclipse of 1900 in Tripoli, see *The Nation*, June 21, 1900, 473–474, and *North Africa*, July 1900, 75–76. The view from the roof of the British consulate is described by Cowper (1897:2–4) and captured by David Todd's photographs, DPTP, Box 45, Folders 316 and 319.
99 "The day was splendid": *The Boston Daily Globe*, May 29, 1900, 5.
99 "Mr. Lowell is better, decidedly": Entry for June 13, 1900, journal of Mabel Loomis Todd, MLTP, Diaries and Journals Microfilm, Reel 9.
100 "This threshold of a new year": *The Times-Democrat* (New Orleans), January 6, 1901, Part Second, 5.
100 fiercely debated for years: "The Herald has received a thousand letters from readers on this subject. For years no controversy seems to have so stirred up the thinking and arguing public." *The New York Herald*, December 31, 1899, Sixth Section, 11.
100 "At midnight of December 31, 1900": *The New York Herald*, December 17, 1899, Fifth Section, 6. Also see Flammarion's comments in *L'Astronomie*, October 1890, 394; *The New-York Times*, February 28, 1892, 17; and *The New York Herald*, January 1, 1900, 3–4.
100 "Let the people of the other worlds": *The Examiner* (San Francisco), December 29, 1900, 2.
100 "watch meetings," held as fundraisers: *The Deseret Evening News* (Salt Lake City), December 22, 1900, 15; *The New York Herald*, December 25, 1900, 11; *Buffalo Courier*, December 30, 1900, 14.
100 Among the celebrities who had sent: *The Chicago Daily Tribune*, January 1, 1901, 1–2 and 18; *The San Francisco Call*, January 1, 1901, 4.
101 "To the American Red Cross": Cheney and Uth (1999:94); *The Examiner* (San Francisco), January 1, 1904, 4; *The World* (New York), January 2, 1901, 8.

CHAPTER 9: MESSENGERS

105 HAS NIKOLA TESLA SPOKEN WITH MARS?: *New York Journal and Advertiser*, January 1, 1901, 4.
105 his laboratory on East Houston Street: Photographs and illustrations of Tesla's laboratory are in Tesla (1899); McGovern (1899); and W. B. Carlson (2013:219).
105 face and hands were smeared with grease: *New Ideas*, February 1901, 1.
105 "I would have abstained": *The Sun* (New York), January 3. 1901, 7.
106 "Their character showed unmistakably": *The Evening Telegram* (New York), January 3, 1901, 10 O'Clock P.M. Edition, 7.
106 "Inhabitants of Mars, I believe": *The Evening Telegram* (New York), January 3, 1901, 10 O'Clock P.M. Edition, 7. See also *The Philadelphia Inquirer*, January 13, 1901, 1. To at least one other reporter, Tesla suggested that the signals might have come from Venus. See *The Indianapolis Journal*, January 2, 1901, 4.

NOTES ON SOURCES

106 **WORLD SPEAKS TO WORLD:** *The Examiner* (San Francisco), January 4, 1901, 1.
106 **TESLA THINKS HE CAN TALK:** *The Philadelphia Inquirer*, January 3, 1901, 1.
106 **WILL THE DAWN OF THE NEW CENTURY:** *The New York Herald*, January 13, 1901, Fifth Section, 4.
107 **merely a Martian cloud:** *The New York Herald*, January 9, 1901, 11.
107 **"Are these lights the result":** *Los Angeles Daily Herald*, January 27, 1901, 8.
107 **"Of this you may be sure":** *New York Journal and Advertiser*, January 9, 1901, 16. Although unsigned, the piece was reprinted in Brisbane (1906:281–284). Brisbane's authorship is revealed by *The Scrap Book*, September 1906, 98.
107 **"All the News That's Fit to Print":** Berger (1951:117).
107 **"I should attribute the alleged signals":** *Daily Mail* (London), January 5, 1901, 3.
108 **"Until Mr. Tesla has shown":** Holden (1901:444).
108 **astronomers dismissed Tesla as a "visionary":** *New York Journal and Advertiser*, January 4, 1901, 2.
108 **"It is intimated by one or two":** *The Chicago Sunday Tribune*, January 13, 1901, 9.
108 **he attempted to stay professional:** Tesla answered his detractors in *The New York Herald*, January 6, 1901, Second Section, 7; *New York Journal and Advertiser*, January 10, 1901, 1; and *The Sun* (New York), January 11, 1901, 6.
108 **"All the tame beasts":** *The North American* (Philadelphia), January 11, 1901, 8.
109 **"Mars!" the boy cried:** *The New Golden Hours*, April 27, 1901, 9.
109 **an astonishing seventy miles per hour:** *St. Louis Post-Dispatch*, November 17, 1901, Part Three, 1.
109 **an age "of promise and hope":** *The World* (New York), December 29, 1900, 1.
109 **"drift through space on flying machines":** *The Examiner* (San Francisco), December 23, 1900, "The Twentieth Century" Sunday Supplement, 8.
109 **"from San Francisco to London":** *New-York Tribune*, January 6, 1901, "The Story of the Nineteenth Century" Sunday Supplement, 8.
109 **WHAT MAY HAPPEN IN THE NEXT HUNDRED YEARS:** *The Ladies' Home Journal*, December 1900, 8.
109 **"The proverb that there is nothing new":** *The San Francisco Call*, January 5, 1901, 9.
110 **"Send us up some Pears' Soap":** *Life*, July 4, 1901, back cover.
110 **"Not feeling well, getting old and weak":** *The Brooklyn Daily Eagle*, January 21, 1901, 4.
110 **"Please send one bottle Bailey's Pure Rye":** *The Times* (Philadelphia), January 16, 1901, 7.
111 **"catchy air, which is taking at a rapid rate":** *The Daily Item* (New Orleans), December 8, 1901, 8. See also *Scranton Tribune*, December 14, 1901, 2; *Buffalo Courier*, November 17, 1901, 13; and *Duluth Evening Herald*, December 28, 1901, Last Edition, 9.
111 *A Peep at Mars*: *Springfield Sunday Republican* (Springfield, MA), March 17, 1901, 11.
111 **thirty-two inches tall:** Reports of her height vary, but see *The New York Times*, November 26, 1919, 9, and Magri (1979).
111 **"Be not afraid. I am with you":** Morris (1901:7).
112 **"Well, now for Mars":** All quotations and actions in this scene come from an original typescript of *A Message from Mars* as performed on Broadway, starring the actor Charles Hawtrey, in 1901, APCR, Scripts Box 69. The dialogue and plot were modified slightly by the time the play was published years later. See Ganthony (1923).
112 **"'Latest observations have revealed'":** In the original script, this line included a small typo, which I have corrected. Confirmation that it was a typo can be found in *The Observatory*, February 1900, 110.
112 **New York's Garrick Theatre:** Morrison (1999:13). The play opened on October 7, 1901. See *The New York Times*, October 8, 1901, 9, and October 13, 1901, Magazine

Supplement, 4; *New-York Tribune*, October 8, 1901, 9; *The Evening Post* (New York), October 8, 1901, 7; and *The New York Herald*, October 8, 1901, 10, and October 20, 1901, Fifth Section, 7.

113 **"Why, it was you, was it not":** McCall (1901:4). The book was written by Mary McNeil Fenollosa. For several years, there was much debate as to the true identity of the pseudonymous author. Some suspected it was Percival Lowell. See *The Sunday Herald* (Boston), June 25, 1905, Magazine Section, 4, and *The Boston Herald*, September 12, 1906, 7.

114 **"I believe that Mars is inhabited":** *The Chicago Daily Tribune*, March 26, 1901, 2.

114 **five outlaws who had escaped:** *The Coconino Sun* (Flagstaff), April 6, 1901, 1; also March 9, 1901, 3.

114 **paleontologist with the U.S. Geological Survey:** Ward (1918:244).

114 **a historian from Cornell, an old friend who taught English literature at Harvard:** *The Coconino Sun* (Flagstaff), June 22, 1901, 3.

114 **asked to give the oration:** *The Coconino Sun* (Flagstaff), July 6, 1901, 1.

114 **"Mr. Lowell's work makes Flagstaff":** *The Sunday Inter Ocean* (Chicago), September 8, 1901, 39.

114 **"I am so much at home here":** Percival Lowell to Elizabeth Lowell Putnam, April 12, 1901, in Strauss (2001:195). Note: I was unable to locate the original letter in the archival collection cited.

114 **"I need to talk to someone who understands":** A. E. Douglass to William H. Pickering, March 8, 1901, AEDP, Box 14.

114 **"I am deeply attached to Mr. Lowell":** A. E. Douglass to William Lowell Putnam, March 12, 1901, AEDP, Box 18, Folder 1 ("Lowelliana").

115 **he waited twenty-four hours:** David F. Brinegar, *A. E. Douglass: A Biography*, 1933 (Rev. 1954), unpublished manuscript, 120, AEDP, Box 177, Folder 6 ("Revised Draft").

115 **Douglass stepped forward and was summarily fired:** The date was August 9, 1901; *The Coconino Sun* (Flagstaff), August 10, 1901, 3. For more on the circumstances surrounding Douglass's abrupt dismissal, see Hoyt (1976:123–125); Webb (1983:48–49); and W. L. Putnam et al. (1994:35–41).

115 **"The events occurred many years ago":** A. E. Douglass to A. Lawrence Lowell, February 24, 1937 (mailed August 11, 1937), AEDP, Box 18, Folder 1 ("Lowelliana").

116 **Sir Robert Ball taught astronomy at Cambridge:** Macpherson (1905:99–107); Chapman (2007); R. Jones (2005); Lightman (2007:397–421).

117 **"A pleasant night it was":** Ball (1915:329). Edward S. Morse's attendance at the dinner is indicated in Morse's list of engagements for November 1901, ESMP, E2, Box 30, Volume 11, "Personal Notices," 1901.

117 **Lowell's *Mars* among his favorite books:** *The Academy*, December 10, 1898, 435.

117 **"paid tribute to the recent observations":** *The Times* (Philadelphia), November 10, 1901, 11. See also *The Philadelphia Inquirer*, November 10, 1901, 8.

117 **"I do believe in the existence":** *New York Journal and Advertiser*, October 29, 1901, 6.

117 **"The canals which astronomers":** *New York Journal and American*, November 24, 1901, American Magazine Supplement; *The Examiner* (San Francisco), December 1, 1901, Sunday Examiner Magazine.

117 **"a convert to the theories advanced":** *The Cincinnati Enquirer*, December 8, 1901, 1.

117 **"Since man began the skies to scan":** *The Boston Daily Globe*, April 1, 1902, 14.

118 **Marconi pressed a telephone receiver to his ear:** For details of Marconi's experiment in Newfoundland, see Baker (1902); McGrath (1902); *Electrical World and Engineer*, December 21, 1901, 1023–1025; *The Times* (London), January 3, 1902, 4; and Raboy (2016:170–176).

CHAPTER 10: "SMALL BOY THEORY"

119 **"Our very eye, the most remarkable":** Marion (1867:24).
119 **"[There are] numerous means that art":** Marion (1867:70).
119 **The true author was Camille Flammarion:** Cotardière and Fuentes (1994:111).
120 **Its caption: "All is vanity":** *Life*, November 27, 1902, 459.
120 **"We cannot assume that what we are able":** Maunder (1894:252).
120 **"the veriest dummy that I ever saw":** George B. Airy to J. Walrond, May 28, 1875, GBAP, CUL MS RGO 6/7 f.335. For more on Maunder, see Kinder (2008) and Macpherson (1905:192–200).
121 **"'Canals' in the sense of being":** Maunder (1895:162–163).
121 **"canaliform illusion":** Coined by Antoniadi (1902:82), the term was soon quoted by others, including Maunder (1903a:353).
121 **Royal Hospital School, which prepared the sons:** Turner (1980).
122 **Pack rats crawled in the attic:** A. L. Lowell (1935:150–151). For more on the Baronial Mansion, see W. L. Putnam et al. (1994:76–79) and Henry Giclas, "Reminiscences of the Lowell Observatory," 31–34, HLGP, Working Papers Series, Box 38. Lowell purchased the music box in 1903; see W. Louise Leonard to V. M. Slipher, January 12, 1903, VMSP, V. M. Slipher Correspondence.
122 **gave the university a portion of the vast inheritance:** *The Boston Herald*, March 30, 1903, 9.
122 **it provided little or no compensation:** Percival Lowell to Henry S. Pritchett, January 12, 1904, LOAC, Copybook 5E, 439.
122 **"Anybody can see Mars now":** Serviss (1903).
123 **"Mars is behaving in satisfactory accord":** Percival Lowell to William Lowell Putnam, February 11, 1903, LELP.
123 **"Guglielmo Marconi announced to-night":** *The New York Times*, December 15, 1901, 1.
123 **"[Marconi] has gained so enormous a step":** *New York Journal and American*, December 16, 1901, 2.
123 **"There is no reason why we should not":** *New York Journal and American*, January 15, 1902, 5.
124 **called the place Wardenclyffe:** *The Port Jefferson Echo*, December 15, 1900, 2, and August 3, 1901, 2; *New-York Tribune*, August 7, 1901, 4; *The Brooklyn Daily Times*, August 20, 1901, 4; *The Brooklyn Daily Eagle*, August 11, 1902, 15. Wardenclyffe today is the village of Shoreham, NY.
124 **"There is not one word to say":** *St. Louis Post-Dispatch*, June 10, 1903, 7. For more on Tesla's tower, see *The Brooklyn Daily Eagle*, October 6, 1901, 5, and March 26, 1916, 12; *The Brooklyn Daily Times*, April 10, 1903, 1, and July 16, 1903, 9; *The New York Press*, February 23, 1903, 1; *Daily Star* (Brooklyn), July 18, 1903, 1; *New-York Tribune*, July 19, 1903, 2; *The New York Times*, March 27, 1904, 12; and Tesla (1919b).
124 **"All sorts of lightning were flashed":** *The Sun* (New York), July 17, 1903, 1.
124 **"The villagers sit out in front":** *The World* (New York), July 17, 1903, Evening Edition, 5.
125 **"At a recent meeting of the Astronomical Society in London":** *The World* (New York), July 23, 1903, Evening Edition, 10.
126 **"putting in all that they could see":** All quotations from the experiment are in Maunder (1904b). See additional details in Evans and Maunder (1903); Maunder (1903b); Maunder (1913:107–108); *Knowledge*, November 1902, 251; *Journal of the British Astronomical Association* 13 (1902–1903): 13–14 and 333–338; and *The Standard* (London), December 12, 1903, 7.
127 **"[What] a great pity":** *Journal of the British Astronomical Association* 13 (1902–1903): 335.

127 **"These experiments prove nothing whatever"**: *The English Mechanic*, June 26, 1903, 440. See also *Journal of the British Astronomical Association* 13 (1902–1903): 336–337. Lowell's defender would later write a science-fiction novel and dedicate it to Lowell. Wicks (1911).
128 **"really cut away the ground"**: *Journal of the British Astronomical Association*, 13 (1902–1903): 338.
128 **"It does not seem . . . that the results"**: *The Sunday Times* (Minneapolis), November 15, 1903, Magazine Section, 2.
128 **"Under certain circumstances it is quite possible"**: P. Lowell (1906:202–203).
128 **"Because a small boy would certainly not"**: P. Lowell (1905a:98).
128 **what he derided as Maunder's "Small Boy Theory"**: P. Lowell (1905a:98); P. Lowell (1906:202). Lowell borrowed this epithet from the astronomy writer Agnes Clerke, who colorfully employed it when she spoke to him in London. See diary entry for September 24, 1904, PLD.
128 **"It is not me who makes lines"**: Percival Lowell to Camille Flammarion, January 31, 1908, CFP (original); LOAC, Copybook 9, 85 (letterpress copy).
129 **"The theory expressed by Mr. Maunder"**: *The Pittsburg Press*, September 18, 1904, 5.

CHAPTER 11: THE MARTIANS OF EARTH

130 **"Listen, ye mortals of Earth"**: *New-York Tribune*, April 27, 1903, 1; *The New York Press*, April 27, 1903, 1.
131 **"inclined to believe that the Entoans"**: Weiss (1905:475–476). The book was originally published in 1903. See *St. Louis Post-Dispatch*, December 20, 1903, Sunday Magazine, 8; *The Sunflower* (Lily Dale, NY), February 6, 1904, 1; Hyslop (1913); and Crossley (2011:140–141).
131 **A man from Boston said his soul**: *The World* (New York), December 5, 1897, 43.
131 **"with a slight foreign accent"**: *Borderland*, October 1897, 407. For more on psychic encounters with Mars and Martians, see *The Evening Telegram* (New York), March 18, 1904, 5; *The Sun* (Baltimore), August 20, 1906, 9; *Cleveland Plain Dealer*, July 26, 1908, Sunday Magazine, 4; Owen (2004:157–161); and Davenhall (2018).
131 **"Telepathy is nature's wireless telegraphy"**: *Deseret Evening News* (Salt Lake City), April 8, 1905, 19.
132 **including Flammarion and Schiaparelli**: C. Flammarion (1907); *The New York Herald* (European Edition, Paris), October 9, 1892, 4; *The Sun* (Baltimore), March 3, 1893, 4.
132 ***"Ané éni ké érédute"***: Flournoy (1994:140).
133 **"The order of the words"**: Flournoy (1994:159).
133 ***Métiche* was Martian for *monsieur***: Flournoy (1994:97).
133 **"in spite of its strange appearance"**: Flournoy (1994:135).
133 **nothing more than "disguised French"**: Flournoy (1994:158).
133 **"romance of the subliminal imagination"**: Flournoy (1994:13).
133 **"terrible power—a power"**: Mousseau (1973:56).
134 **"In any active brain ideas"**: Percival Lowell, "Address: Commemoration Day, Flagstaff, Arizona," 10, PLP, Unpublished Lectures, Manuscripts, and Poems, Box 1U, Folder 11. See also *The Coconino Sun* (Flagstaff), May 30, 1903, 3.
134 **"He has become more certain"**: Basil Hall Chamberlain to Lafcadio Hearn, August 5, 1893, in Koizumi (1936:34).
134 **"I cannot stand . . . the way he has"**: Basil Hall Chamberlain to Lafcadio Hearn, January 10, 1893, in Koizumi (1936:2).
134 **"selects and bends facts"**: Basil Hall Chamberlain to Lafcadio Hearn, June 23, 1893, in Koizumi (1937:81).

134 **"It is better never to admit":** A. E. Douglass, handwritten memo dated 1/7/12 and titled "Dr. Lowell's Scientific Standing," AEDP, Box 18, Folder 1 ("Lowelliana"). See also David F. Brinegar, *A. E. Douglass: A Biography*, 1933 (Rev. 1954), unpublished manuscript, 73, AEDP, Box 177, Folder 5 ("Revised Draft").

135 **"It is easily conceivable that a limited water supply":** P. Lowell (1904*b*:43).

136 **"Each [canal] waxed and waned":** P. Lowell (1904*a*:62).

137 **51 miles per day, or 2.1 miles per hour:** P. Lowell (1904*a*:85).

137 **"The mental ear detects the sound":** P. Lowell (1908:198).

137 **"I think I have found the law governing":** Percival Lowell to V. M. Slipher, November 10, 1903, LOAC, Copybook 5E, 401–402.

137 **"from the north pole to the equator":** Percival Lowell to C. O. Lampland, November 10, 1903, LOAC, Copybook 5E, 398.

137 **"Since there is no force to make water":** Percival Lowell to V. M. Slipher, November 10, 1903, LOAC, Copybook 5E, 403.

138 **flirted for a time with a Japanese geisha:** Greenslet (1946:351–354).

138 **Lowell mentioned her several times:** P. Lowell (1886:244–249, 370).

138 **"Lowell ought to know something about the women":** George C. Foulk to "My dear Parents and Brothers," March 10, 1886, addendum dated March 13, GCFP Microfilm; also in Hawley (2008:150).

138 **close relationship with his longtime secretary:** Hollis (1992); W. L. Putnam et al. (1994:69–86). After Lowell's death, his secretary published a tribute to her late boss. Leonard (1921).

139 **"the young woman's clothing draped":** David F. Brinegar, *A. E. Douglass: A Biography*, 1933 (Rev. 1954), unpublished manuscript, 131, AEDP, Box 177, Folder 6 ("Revised Draft").

139 **"When will her Ladyship be at home":** Percival Lowell [to Irva Struthers], dated "Sunday," on stationery from the Hotel Manhattan, Scrapbook "The Story of a Friendship," PLP, Personal, Box 3.

139 **He first spied her at tea:** Details of their relationship can be found in an unpublished remembrance by Irva Struthers McCall, "In memoriam: P.L. 1916," PLP, Personal, Box 3.

139 **well traveled, artistic:** *The Philadelphia Inquirer*, October 6, 1896, 6, and November 9, 1900, 8; *The New York Times*, November 4, 1902, 9.

139 **"To the missed of Earth / The mist of heaven":** Percival Lowell [to Irva Struthers], September 1903, Scrapbook "The Story of a Friendship," PLP, Personal, Box 3.

139 **he a nonbeliever:** Greenslet (1946:366); Basil Hall Chamberlain to Lafcadio Hearn, August 27, 1893, in Koizumi (1937:97).

139 **"Alone! yes, always alone":** Percival Lowell [to Irva Struthers], n.d., on stationery from the Hotel Walton, Scrapbook "The Story of a Friendship," PLP, Personal, Box 3.

139 **Struthers accepted the hand of another:** *The Philadelphia Inquirer*, July 17, 1904, 8, and December 4, 1904, Third Section, 1.

139 **"And so I go back again":** Percival Lowell [to Irva Struthers], n.d., on stationery from the Hotel Walton, Scrapbook "The Story of a Friendship," PLP, Personal, Box 3. Lowell used much the same language a year later: "I miss the only real solace of life, work." Diary entry for September 5, 1904, PLD.

CHAPTER 12: TRUTH IN THE NEGATIVE

141 **"An amusing fact occurred to me":** Percival Lowell to Camille Flammarion, February 13, 1905, CFP.

141 **Lowell ventured out by tram and foot:** Diary entry for August 7, 1904, PLD.

141 **"Your theory of vegetation":** Giovanni Schiaparelli to Percival Lowell, Decem-

NOTES ON SOURCES 273

ber 4, 1904, in Morse (1906:62). The original letter received by Lowell has gone missing, but Schiaparelli's draft is at the Brera Observatory [GSP] and is transcribed and translated in J. Putnam and W. Sheehan (2021:197–198).
142 **"It was an agreeable table"**: Diary entry for October 2, 1904, PLD.
142 **"The latest observations of Percival Lowell"**: Serviss (1904:21).
143 **"We wish to publish your despatches"**: C. V. Van Anda to Percival Lowell, April 19, 1905, PLP, Percival Lowell Correspondence.
143 **Carr Van Anda:** Van Anda was revered for his discerning news judgment. "He scents buncombe and fraud miles away," wrote one admirer in *The Editor and Publisher*, May 26, 1917, 18. See more at Fine (1933).
143 **"an excursion into fairyland"**: Maunder (1904a:88).
143 **"Mars, unfortunately, does not lend itself"**: Maunder (1904a:89).
143 **"No photograph yet taken of Mars"**: Clipping headlined THE CANALS OF MARS, source unidentified, February 1904, PLP, Lowell Scrapbook 1894–1906.
144 **Martian surface was "impossible"**: P. Lowell (1895b:644).
144 **Lampland experimented with color filters:** Lampland (1905); P. Lowell (1906:274–277); *The Boston Herald*, February 1, 1906, 5; Hoyt (1976:175–177).
144 **"How gets on the photographing"**: Percival Lowell to C. O. Lampland, May 16, 1904, PLP, Percival Lowell Correspondence.
145 **Geronimo, the conquered Apache leader:** Parezo and Fowler (2007:95–96, 109–115).
145 **Garrett P. Serviss Jr.:** *The Brooklyn Citizen*, August 30, 1904, Last Edition, 6; Lucas (1905:38–39).
145 **"Helen Keller's address at the Hall of Congresses"**: *The St. Louis Republic*, October 23, 1904, Part 4, 1.
145 **"One may ask how we can possibly"**: *The Duluth News Tribune*, January 13, 1905, 4.
146 **"It will convert the entire earth"**: *The Boston Sunday Globe*, December 18, 1904, Magazine Section, 3. See also Tesla (1904).
146 **dismissed as the delusions of a hollow romantic:** Hawkins (1903).
146 **"My first 'world telegraphy' plant"**: Tesla (1905:24).
147 **"My enemies have been so successful"**: Nikola Tesla to J. P. Morgan, December 11, 1903, NTC, Microfilm Reel 3.
147 **"You are a big man, but your work"**: Nikola Tesla to J. P. Morgan, December 19, 1904, NTC, Microfilm Reel 3.
147 **"Let me tell you once more"**: Nikola Tesla to J. P. Morgan, February 17, 1905, NTC, Microfilm Reel 3.
147 **exacerbated by romantic disappointment:** W. B. Carlson (2013:361–362).
147 **"Sometimes we feel so lonely"**: *The Evening Star* (Washington, D.C.), June 6, 1896, 15.
147 **signed his name: "Nikola Busted"**: Nikola Tesla to Robert U. Johnson, January 24, 1904, NTP, Microfilm Reel 1.
147 **"Canals already seen"**: Percival Lowell to Edward S. Morse, January 15, 1905, ESMP, E2, Box 9, Folder 1.
147 **Lowell now invited Morse:** Wayman (1942:394).
148 **"for the doubles do not come on"**: Percival Lowell to Edward S. Morse, May 4, 1905, ESMP, E2, Box 9, Folder 1.
148 **"the resemblance of its lambent saffron"**: P. Lowell (1906:149).
148 **"Pitiless as our deserts are"**: P. Lowell (1906:157).
148 **"The ant stands among the invertebrates"**: Morse (1906:156).
148 **"Am in despair of seeing anything"**: Morse (1906:162).
148 **"I . . . have come to the conclusion"**: Morse (1906:163).
148 **"Enigmatic functional alternation"**: Percival Lowell to Edward C. Pickering, March 9, 1905, PLP, Percival Lowell Correspondence.
149 **"First winter snow . . . in arctic region"**: Percival Lowell to Edward C. Pickering, May 20, 1905, PLP, Percival Lowell Correspondence.

149 **CAMBRIDGE, Mass., May 27**: *The New York Times*, May 28, 1905, 1.
149 **"To-day we can state as positive"**: *The Sun* (New York), July 16, 1905, Second Section, 7; also *The Washington Post*, July 16, 1905, 8. Some of the quotations attributed to Lowell in this article do not ring true. It seems likely that an enterprising journalist took the contents of his recently published *Lowell Observatory Bulletin No. 21* [P. Lowell (1905b)] and recast its contents as if from an interview.
150 **"hardly larger than the head of a shirt-stud"**: *The Morning Post* (London), December 1, 1905, 10.
150 **"Here they are," Flammarion said**: *Le Petit Parisien*, October 26, 1905, 2.
151 **"To make them out would have required"**: Selley (1907:242).
151 **"Personally I find it extremely difficult"**: *The Observatory*, August 1905, 336.
151 **"The canals were clear and unmistakable"**: *The Journal of the British Astronomical Association* 16 (1905–1906): 5.
151 **"Dear fellow-Martian"**: Percival Lowell to Edward S. Morse, June 29, 1905, ESMP, E2, Box 9, Folder 1.

CHAPTER 13: PLANET OF PEACE

152 **"Plans have been completed for assembling"**: McGee (1904:4).
153 **"McGee was an inexhaustible mine"**: Harris (2001:255).
153 **"The Americans treat us"**: Parezo and Fowler (2007:274). For more on the Anthropology Villages, see Breitbart (1997).
153 **"that innate and intuitive curiosity"**: McGee (1904:4).
153 **artist's rendition of possible Martian life-forms**: *New York American* and *Hearst's Boston American*, March 11, 1906, and *The San Francisco Examiner*, March 18, 1906, all in American Magazine Section, 6. Similar illustrations ran two years earlier in *The Washington Times* and *The Boston Sunday Journal*, April 17, 1904, Colored Section, 10.
153 **"I cannot understand why people persist"**: *The Sunday Star* (Washington, D.C.), January 28, 1906, Editorial Part, 1. This point is echoed by Robinson (1908) and *San Francisco Chronicle*, June 18, 1909, 6.
154 **"Gravity on the surface of Mars"**: P. Lowell (1895a:202–203). Kaempffert (1907:486) makes a similar argument.
154 **"At first it might be supposed"**: Ball (1894:727). See also Garrett P. Serviss in *The San Francisco Examiner*, April 7, 1907, Editorial Section, 2.
155 **gave Lowell's theories favorable coverage**: Markley (2005:105–109).
155 **"Whatever atmosphere exists on Mars"**: Gregory (1900:921). See also David Todd in *The New York Herald*, October 27, 1907, Third Section, 5.
155 **"the man on Mars [may] have an elephantine nose"**: *The Philadelphia Inquirer*, November 23, 1919, Feature Section, 3.
155 **"very big eyes—wonderful eyes"**: *The World* (New York), July 31, 1892, 6.
156 **"precisely the kinds [of creatures]"**: Morse (1906:149). Another American zoologist, William K. Brooks, also suggested that the Martians might be arthropods. *New York American*, December 13, 1908, Editorial Section, 5.
156 **PROF. E. S. MORSE THINKS [THE MARTIANS] MAY BE GIANT ANTS**: *The Sunday Herald* (Boston), October 21, 1906, Magazine Section, 4.
156 **"Since many of the 'canals' were photographed**: *New York American*, December 2, 1905, 16.
156 **"recognized throughout the world as an authority"**: *The World* (New York), July 8, 1906, Sunday Magazine, 1.
156 **"whose logical deductions [are] as full"**: *The Sunday Herald* (Boston), October 29, 1905, Magazine Section, 8.
156 **puzzled over whether he and Lowell ever met**: Crossley (2011:323).
157 **"He is soon to visit Boston"**: *The Boston Post*, April 16, 1906, 3.
157 **Wells attended an afternoon reception**: *Boston Evening Transcript*, Last Edi-

tion, April 20, 1906, 11. See also *Boston Evening Transcript*, Last Edition, April 25, 1906, 17.
157 **"Why not?" Wells replied:** *The Times Dispatch* (Richmond), May 21, 1906, 3.
157 **"Prof. Lowell told me many things":** *The World* (New York), July 8, 1906, Sunday Magazine, 1.
157 **"You could give us the first governor":** *Albuquerque Evening Citizen*, July 30, 1906, 1.
158 **a notorious controversy over race:** Newkirk (2015).
158 **"We do not like this exhibition":** *The New York Times*, September 11, 1906, 2.
158 **"The first thing that is forced on us":** P. Lowell (1906:377).
158 **"quite simply a small masterpiece":** Giovanni Schiaparelli to Percival Lowell, February 1, 1907, GSP. Transcription and translation in J. Putnam and W. Sheehan (2021:204–206).
158 **"When a planet has attained to the age":** P. Lowell (1906:378).
158 **"[Mars] is like a sinking ship":** *Boston Evening Transcript*, March 31, 1906, Last Edition, Part Three, 3.
159 **as if it were grand opera night:** Leonard (1921:25).
159 **quickly overflowed the auditorium:** "The crowds that gathered for this lecture, a large part of whom could not even get standing room, ought to encourage Mars in his planetary progress." Lilian Whiting, in *The Times-Democrat* (New Orleans), October 21, 1906, Part Third, 8. Whiting continued to praise Lowell's lecture series in later columns for the same newspaper on November 4, November 25, and December 30, 1906, all Part Third, 8.
159 **he had "not been idle":** *Boston Evening Transcript*, October 16, 1906, Last Edition, 16.
160 **"Mars is a vast Saharan world":** *Boston Evening Transcript*, October 19, 1906, Last Edition, 3.
160 **"Not only do the observations":** *Boston Evening Transcript*, November 9, 1906, Last Edition, 12.
160 **He flashed a playful smile:** *Boston Evening Transcript*, September 26, 1907, Last Edition, 10. Lowell's lecturing was described as "charming" (*Boston Evening Transcript*, October 22, 1906, Last Edition, 10); displaying "an enjoyable, easy way" (*The Boston Daily Globe*, October 24, 1906, 7:30 P.M. Edition, 13); having a "racy, epigrammatic brilliancy of style" (*Boston Evening Transcript*, October 31, 1906, Last Edition, 15); and showing a "deep earnestness . . . now and then relieved by a touch of quiet and delicate humor" (*The Boston Sunday Globe*, November 11, 1906, 41).
160 **"Some questioned the facts set forth":** Ritchie (1907:102).
160 **"Well, I call him a visionary!":** *The Boston Herald*, October 31, 1906, 6.
160 **"In many a household":** *The Boston Daily Globe*, October 31, 1906, 7:30 P.M. Edition, 10. See also *The Sunday Post* (Boston), November 18, 1906, 30.
160 **"the great astronomer who is now held":** *The New York Times*, December 9, 1906, Part Four, 1.
161 **WILL THE NEW YEAR SOLVE THE RIDDLE OF MARS?:** *The New York Herald*, December 30, 1906, Magazine Section, 5.

CHAPTER 14: ALIANZA

162 **"[We] can see [the] lively and intelligent":** *The Los Angeles Times*, May 24, 1905, Part 2, 4.
162 **"if they have heads and feet":** *The Evening Star* (Washington, D.C.), May 23, 1905, 4.
162 **"This is a real taste of winter":** Clipping that begins "I met Prof Percival Lowell," source unidentified but published in February 1907 (which suggests that the

NOTES ON SOURCES

writer's encounter with Lowell occurred during the great blizzard of February 5), PLP, Lowell Scrapbook 1906–1911.

163 **Reporters shouted and news cameras flashed:** Front page of *The Boston Post*, *The Boston Daily Globe*, *The Boston Herald*, and *Boston Daily Advertiser*, November 28, 1906. Also, on the same date: *The Sun* (New York), 8, and *The Boston Journal*, 12.

163 **"No one who is interested in the wonderful development":** *The Semi-Weekly Times-Democrat* (New Orleans), July 18, 1905, 5.

163 **"that gaunt frozen border land":** Peary (1907:51).

163 **"There is no higher, purer field":** International Geographical Congress (1905:79).

163 **3 degrees shy of his goal (or so he claimed):** Herbert (1989:184–185).

163 **"Between those two great cosmic boundaries":** Peary (1907:xi). See also *The Washington Post*, December 16, 1906, 1.

164 **conversed at length about matters scientific:** *The Boston Herald*, November 28, 1906, 1. Same date, see also *Boston Evening Transcript*, Last Edition, 17, and *The Boston Traveler*, 1.

164 **Ends of the Earth Club:** Edward S. Morse also belonged to the club. See membership list in "The Ends of the Earth 1906," ESMP, E2, Box 31, Volume 12.

164 **compared Peary's daring trek to "a journey to Mars":** *The Semi-Weekly Times-Democrat* (New Orleans), July 18, 1905, 5.

164 **"less appealing to the gallery":** P. Lowell (1906:54).

165 **"The riddle of the planet Mars":** *The Washington Post*, May 13, 1907, 1.

165 **he made a proposal to his wealthy friend:** D. Todd (1920:49); D. Todd (1908a:73); *The Evening Star* (Washington, D.C.), December 31, 1907, 9; Percival Lowell to C. O. Lampland, February 2, 1907, LOAC, Copybook 7, 488. The letter of agreement stipulated that the cost to Lowell would be $3,500. "Lowell Observatory Expedition to Peru," David Todd, February 28, 1907, ECSP, E. C. Slipher Correspondence.

165 **"I hope that [this campaign] is":** Giovanni Schiaparelli to Percival Lowell [n.d., but in reply to letter from Lowell circa March 1907], GSP. Transcription and alternate translation in J. Putnam and W. Sheehan (2021:206–207). Also see comment by Schiaparelli in *The Daily Telegraph* (London), July 10, 1907, 12.

166 **"How exciting if this year":** Camille Flammarion to [Percival Lowell], January 10, 1907, ESMP, E2, Box 4, Folder 6. The letter was addressed merely to "Dear colleague and friend," but the recipient was clearly Lowell, for the note was in reply to Lowell's letter to Flammarion dated November 30, 1906, LOAC, Copybook 7, 386. Lowell had asked Flammarion to induct Edward S. Morse into the Astronomical Society of France. Flammarion's reply said he acceded to the request. Lowell must have shared the letter with Morse, which is how it found its way into Morse's papers.

166 **"This will be the largest telescope":** *The Hartford Courant*, May 13, 1907, 11.

166 **"She will be the Mrs. Peary":** *New York American*, May 11, 1907, 7.

166 **"Don't forget tooth brush and telescope":** Percival Lowell to David Todd, May 11, 1907, DPTP, Box 52, Folder 380.

166 **"Curiously enough 'tooth-brush' was forgotten":** David Todd to Percival Lowell, May 16, 1907, PLP, Percival Lowell Correspondence.

166 **packed in 119 boxes:** *Elizabeth Daily Journal* (Elizabeth, N.J.), March 20, 1908, 21; *Chicago Examiner*, March 1, 1909, 7.

167 **"Almost the only difference":** D. Todd (1908b:346).

167 **"Before we dig any farther on that canal":** *The Minneapolis Journal*, July 8, 1905, 4.

168 **impressed by the visit to Culebra:** "And the great Culebra cut! It is a titanic work. The steam shovels, the blasting, the digging, the trains puffing hither and yon carrying 'dump,' the machine shops in full swing, the chimneys and smoke

and noise—all set in the midst of that dreamy, tropical landscape is positively thrilling in its terrific energy and forcefulness of achievement." Entry for May 22, 1907, journal of Mabel Loomis Todd, 196, MLTP, Diaries and Journals Microfilm, Reel 9.

168 **"Have seen a lot of this terrene canal":** David Todd to Percival Lowell, May 20, 1907, PLP, Percival Lowell Correspondence.
168 **British-dominated mining industry:** Blakemore (1974).
168 **"A trip to the moon, or even Mars":** Entry for June 17, 1907, diary of David Todd, DPTP, Box 52, Folder 382. Many details of the journey through Chile also come from Mabel Loomis Todd's journal (MLTP, Diaries and Journals Microfilm, Reel 9) and diary (MLTP, Diaries and Journals Microfilm, Reel 5), as well as the diary of the Todds' daughter, Millicent, who joined them later in the expedition (MTBP, Box 128, Folders 9–10). See also D. Todd (1908*b*); M. L. Todd (1907*a*); and Mabel Loomis Todd's unpublished "The Nitrate Wealth of Tarapacá," MLTP, Box 79, Folder 350.
168 **largest of these outposts, Oficina Alianza:** Cuthbert (2011); Purbrick (2017).
169 **A loaf of bread might cost a peso:** Lafertte (2014:242).
169 **"I have failed, so far":** Entry for June 26, 1907, journal of Mabel Loomis Todd, 257, MLTP, Diaries and Journals Microfilm, Reel 9.
169 **"Through all these clear and wonderful nights":** Entry for June 26, 1907, journal of Mabel Loomis Todd, 257, MLTP, Diaries and Journals Microfilm, Reel 9.
170 **"It is awe-inspiring and mighty":** Entry for July 2, 1907, diary of Mabel Loomis Todd, MLTP, Diaries and Journals Microfilm, Reel 5. See also M. L. Todd (1907*b*:265); M. L. Todd (1908:792); and *Elizabeth Daily Journal* (Elizabeth, NJ), March 20, 1908, 21.
170 **French sardines and Roquefort cheese:** Percival Lowell to H. Jevne, April 29, June 7 and 27, and July 9, 1907, LOAC, Copybook 8; 135, 207, 234, and 241.
170 **claret and cigars:** Wrexie Louise Leonard to Kenneth MacDonald, April 8 and 13, and May 7, 1907, LOAC, Copybook 8; 77, 90, and 157.
170 ***The New York Times* once again telegraphed:** *The New York Times* to Percival Lowell, July 7 [1907], PLP, Percival Lowell Correspondence.
170 **"All the world is patiently waiting":** *The New York Times*, July 8, 1907, 6.
170 **"Has the secret of life on Mars":** *Boston American*, July 24, 1907, 4.
170 **Douglass used his new platform to proclaim himself an anti-canalist:** Douglass (1907*a*); Douglass (1907*b*).
171 **"Professor A. E. Douglass, who":** *The English Mechanic*, July 12, 1907, 534.
171 **fired "for untrustworthiness":** Percival Lowell to William D. McPherson, June 12, 1907, LOAC, Copybook 8, 216.
171 **negotiated as a quid pro quo:** Strauss (2001:209).
171 **hundreds of pictures of Mars each night:** Slipher (1907). Also see interview with Albert Ilse, who assisted with the photography, DPTP, Box 53, Folder 390.
171 **he established a cypher code:** Percival Lowell sent David Todd a three-page list of code words, DPTP, Box 52, Folder 379.
171 **Mabel Todd awoke at 6:30 A.M.:** Details of this day come from Mabel Loomis Todd's diary entry for July 4, 1907 (MLTP, Diaries and Journals Microfilm, Reel 5), and her journal entry for July 9, 1907, 271–272 (MLTP, Diaries and Journals Microfilm, Reel 9).
172 **NIL SPOT DOUBLE FEEBLE:** David Todd to Percival Lowell, July 4, 1907, PLP, Percival Lowell Correspondence. The telegram opened with an additional code word—FOUR—which meant that Todd judged the atmospheric conditions to be very good for astronomical work, rating them a 4 on a scale of 1 to 5.
172 **"The results have proved":** *The New York Times*, July 9, 1907, 7.
172 **ten thousand photographs of Mars:** David Todd at first estimated that seven

thousand Mars photographs had been taken, but the final count exceeded thirteen thousand. See Slipher (1914:156).

172 **"As they looked at these photographs":** David Todd to Percival Lowell, August 8, 1907, PLP, Percival Lowell Correspondence.

172 **"Bravo!" came Lowell's salute:** Percival Lowell to David Todd, July 26, 1907, DPTP, Box 52, Folder 381; also LOAC, Copybook 8, 272.

172 **Mumm's Extra Dry Champagne:** Percival Lowell to H. Jevne, August 12, 1907, LOAC, Copybook 8, 320.

173 **carriages and motorcars lined Fifth Avenue:** *The New York Times*, October 23, 1907, 2–3; *New-York Tribune*, October 23, 1907, 1–2 and 8.

173 **"the financial markets are in a state of distress":** *The Wall Street Journal*, October 23, 1907, 1. For more on the Panic of 1907, see Bruner and Carr (2023).

173 **chased them down in the city:** "Reporters stand in line, call on us at all hours, and beseech photographs of Mars." Entry for October 24, 1907, diary of Mabel Loomis Todd, MLTP, Diaries and Journals Microfilm, Reel 5.

173 **"What sort of people live on Mars?":** *Boston American*, October 27, 1907, 45.

173 **"The camera has not discovered":** *The New York Times*, October 27, 1907, Magazine Section, 3.

173 **"I suppose the thing most interesting":** *The Sun* (New York), October 24, 1907, 9.

174 **"I have proved," he said:** *Boston American*, October 27, 1907, 45. See also *The New York Herald*, October 27, 1907, Third Section, 5.

174 **DOUBTS ARE ALL DISPELLED:** *Springfield Daily Republican* (Springfield, MA), October 25, 1907, 5.

174 **PEOPLE IN MARS? SURE THERE ARE:** *The Evening Nonpareil* (Council Bluffs, IA), November 1, 1907, 6.

174 **MARS PROVEN INHABITED:** *Alaska Daily Record* (Juneau), October 28, 1907, 4.

174 **"I have your telegram":** Percival Lowell to "Editor of the New York Times," September 7, 1907, LOAC, Copybook 8, 381.

174 **sent draft articles to experts:** John (1981:118).

174 **"The article is very interesting":** Nikola Tesla to Robert U. Johnson, October 3, 1907, NTP, Microfilm Reel 1. On the two men's relationship, see R. U. Johnson (1923:399–401).

175 **sketched invisible notes on the tablecloth:** *The Los Angeles Times*, August 31, 1907, Part 2, 6.

175 **"Tesla has not appeared in the newspapers":** *The Indianapolis Star*, June 27, 1907, 8.

175 **"Signalling to that planet":** Tesla (1907:120).

176 **"Now, the importance of these":** P. Lowell (1907:309).

177 **now prominently displayed at MIT:** *The Sunday Herald* (Boston), December 1, 1907, 4; *The Washington Post*, March 23, 1908, 12.

177 **an invitation to lecture about Mars:** National Geographic Society to Percival Lowell, September 23, 1907, PLP, Percival Lowell Correspondence.

177 **"Look back upon the year 1907":** *The Wall Street Journal*, December 28, 1907, 1. PLP, Lowell Scrapbook 1906–1911.

CHAPTER 15: ARTICLES OF FAITH

183 **"Men, it has been well said":** Mackay (1841:3).

183 **"4600 florins, a new carriage":** Mackay (1841:142).

184 **"On Mars we can . . . see":** Ward (1907:164).

184 **"Since reading your book":** G. R. Wieland to Percival Lowell, February 19, 1907, PLP, Percival Lowell Correspondence.

184 **"almost beyond proof":** *The Pittsburg Press*, August 18, 1907, 8.

184 **"To us, it is a joke":** Edward C. Pickering, "Reading of resolution, and abstracts

of letters re Mars," in typed minutes of meeting at Yerkes Observatory, August 21, 1909, RAAS, Subgroup 2, Box 6, Folder 8.
185 **"All national institutions of churches"**: Paine (1795:4).
185 **a "strange conceit"**: Paine (1795:42–43).
185 **"Let us suppose that communication"**: *Blue Grass Blade* (Lexington, KY), May 2, 1909, 10.
185 **"by enlarging . . . the numbers"**: Selley (1907:253).
185 **"The gracious God that guides"**: *The New York Herald*, August 30, 1908, Magazine Section, 7.
185 **"We imagine one of the first questions"**: *The Christian Work and the Evangelist*, August 3, 1907, 134.
186 **"What on earth can rival the importance"**: *The Kansas City Times*, September 16, 1907, 2.
186 **"We most of us believe . . . that Mars"**: *Sunday World-Herald* (Omaha), February 9, 1908, Editorial Section, 2.
186 **soon released in book form:** P. Lowell (1908).
186 **"Their presence certainly ousts us"**: *The Indianapolis Star*, May 27, 1908, 6.
186 **Mr. Skygack, from Mars:** Holtz (2012:274).
187 **PAIR WAS PROBABLY GUILTY:** *The Pittsburg Press*, February 24, 1908, 6.
187 **"I shall consider the subject"**: *Daily People* (New York), March 22, 1908, 5.
187 **"What sort of inhabitants may Mars possess?":** Wells (1908:335).
187 **"They will probably have heads and eyes"**: Wells (1908:340–342).
188 **"The woman's club movement"**: Gilman (1905:164). See more at Bailey (1905) and Rogers (1905).
189 **"Mabel Loomis Todd . . . is being idolized"**: *Boston Evening Record*, March 4, 1908, 4.
189 **"We want The Latest News from Mars"**: Emily C. Morton to Mabel Loomis Todd, November 13 [1907], MLTP, Box 59, Folder 150.
189 **more than 150 lectures:** Mabel Todd's speaking-engagement notebooks list 152 lectures from November 1907 to December 1909. Of those, 79 were about Mars. MLTP, Box 57.
189 **"That first glimpse of Mars!":** *Elizabeth Daily Journal* (Elizabeth, NJ), March 20, 1908, 21.
189 **five hundred people showed up:** Entry for March 18, 1908, diary of Mabel Loomis Todd, MLTP, Diaries and Journals Microfilm, Reel 5.
190 **"We went down to South America"**: *The Providence Daily Journal*, March 10, 1908, 3; *The Evening Bulletin* (Providence), March 10, 1908, 4.
190 **"Talked on Mars till their hair"**: Entry for March 9, 1908, diary of Mabel Loomis Todd, MLTP, Diaries and Journals Microfilm, Reel 5.
190 **"Took them off their feet:"** Entry for February 25, 1909, diary of Mabel Loomis Todd, MLTP, Diaries and Journals Microfilm, Reel 5.
190 **"Room packed almost to breathlessness"**: Entry for January 15, 1908, diary of Mabel Loomis Todd, MLTP, Diaries and Journals Microfilm, Reel 5.
190 **"So completely had she"**: *Evening Express* (Portland, ME), February 14, 1908, 14.
190 **"The eternal silence of these"**: Pascal (1962:221).
190 **the Olympic athlete, died at twenty-six:** *Democrat and Chronicle* (Rochester, NY), December 25, 1907, 3.
190 **following his own mother:** *The New York Times*, January 20, 1906, 9.
190 **"Disastrous earthquakes . . . have"**: *The New York Herald* (European Edition, Paris), February 18, 1907, 9.
191 **"Do you think there will ever"**: *The Evening Star* (Washington, D.C.), January 10, 1908, 18.
191 **"Mars will be ten percent nearer"**: *The Boston American*, May 2, 1909, 8S.

NOTES ON SOURCES

191 **SOME QUESTIONS MARS MIGHT ANSWER:** *Syracuse Journal*, June 17, 1909, Evening Edition, 8; *The Pittsburg Press*, June 13, 1909, Editorial Section, 4; *The Omaha Daily News*, June 13, 1909, 2D.
191 **"I expect to find an answer":** *The Sunday Post* (Boston), May 16, 1909, 25.

CHAPTER 16: SKYWARD

192 **"Any man who is a bear":** *The New York Times*, December 11, 1908, 1.
192 **"Everything, anything, is possible":** *The New York Times*, October 11, 1908, Magazine Section, 6.
193 **WRIGHT CONQUERS AIR:** *New-York Tribune*, August 9, 1908, 1.
193 **"They treat you in France":** Milton Wright to Wilbur Wright, September 9, 1908, WWOWP, Box 5, Folder 11.
193 **"Ever since the Wrights came back":** *The San Francisco Examiner*, May 19, 1909, 20.
193 **Charles S. Rolls, a swashbuckling aristocrat:** *The Times* (London), July 13, 1910, 12.
194 **Abbott Lawrence Rotch, a prominent Bostonian:** *American Magazine of Aeronautics*, December 1907, 6; *Science*, May 24, 1912, 808–811.
194 **Boston's "Woman Contractor":** *The New York Times*, June 12, 1908, 7. See more at *Boston Evening Record*, April 13, 1908, 4; *The Boston Daily Globe*, June 11, 1908, 1 and 8; and *The Sunday Herald* (Boston), November 1, 1908, Magazine Section, 7.
194 **"Prof. Percival Lowell has married":** *The Chicago Daily Tribune*, June 13, 1908, 10.
194 **"to learn more of Earth":** Percival Lowell to W. L. Leonard, August 18, 1908, in Leonard (1921:34–37). On Lowell's visits to European observatories, see Percival Lowell to V. M. Slipher, August 6 and September 13, 1908, PLP, Percival Lowell Correspondence.
194 **Piccadilly hotel near Buckingham Palace:** The Lowells lodged at Princes' Hotel (Percival Lowell to V. M. Slipher, September 26, 1908, PLP, Percival Lowell Correspondence), "a high-class family hotel" (Baedeker [1908:3]).
195 **the spectators, comprising scientists:** *The Sunday Post* (Boston), September 27, 1908, 36.
195 **the craft launched:** *Daily Mail* (London), September 21, 1908, 3; *The New York Times*, September 21, 1908, 4.
195 **No record survives of Lowell's reactions:** Although Lowell apparently did not record on paper what he experienced during his balloon voyage, others who sailed over London in that era did so. Their writings and aerial photographs allowed me to imagine Lowell's journey. See G. Bacon (1898) and Talbot (1901).
195 **streetlights off and on in unison:** *The Pall Mall Gazette* (London), August 18, 1892, 2.
195 **"The chief function of genius":** P. Lowell (1909:181).
196 **rushing them into print in his latest book:** P. Lowell (1908).
197 **"We must surely account":** *The Christian Register*, September 24, 1908, 1046.
197 **"long since have mastered":** *The New York Herald*, August 15, 1909, Magazine Section, 4.
198 **"The demand of the hour":** *The Times-Democrat* (New Orleans), May 16, 1909, Part Third, 6.
198 **"Let these signals be repeated":** *Collier's*, September 25, 1909, 27–28.
198 **"A large black spot":** *The Evening Star* (Washington, D.C.), April 29, 1909, 15.
198 **"The signaling by means of mirrors":** *The New York Times*, April 25, 1909, Magazine Section, 1.
198 **"Looking down from Mars":** *The Sunday Post* (Boston), April 18, 1909, 27. See more at *The New York Herald*, April 25, 1909, Third Section, 3, and *The Sunday Herald* (Boston), May 16, 1909, Magazine Section, 4.

199 **Texans just might take up the cause:** *The Boston Daily Globe*, April 27, 1909, 7:30 P.M. Edition, 9; *The Washington Post*, April 28, 1909, 5; *St. Louis Post-Dispatch*, April 28, 1909, Seventh Edition, 1; *The Sunday Post* (Boston), May 2, 1909, 21. Pickering later discovered that the funding offer had been nothing more than "an advertising scheme." William H. Pickering, "Remarks on the President's Suggestion Regarding Mars," dated July 7, 1909, enclosed in a letter to William J. Hussey, July 13, 1909, RAAS, Subgroup 2, Box 4, Folder 9.

199 **luncheon with the Wright brothers:** *The Washington Post*, June 12, 1909, 10. For more on David Todd's interest in aviation, see R. O. Jones (1994) and *The Sunday Post* (Boston), May 15, 1910, 44.

200 *If ten miles would cut much ice:* New-York *Tribune*, May 25, 1909, 7.

200 **"Ten miles is higher than man has ever been":** *New-York Tribune*, May 25, 1909, 7.

200 **One of the two highest balloon ascensions:** R. Holmes (2013:211–219); Glaisher et al. (1871:50–58).

200 **Four decades later, in 1901:** Labitzke and van Loon (1999:4–8); Hildebrandt (1908:275–278).

200 **"we shall come down alive":** *New-York Tribune*, May 25, 1909, 7.

200 **"We will have little observation windows":** *New York Evening Journal*, June 17, 1909, 10. See more at *The Denver Post*, May 16, 1909, 3, and *The Sunday Post* (Boston), May 16, 1909, 25.

200 **"If either of the tanks should give way":** *The New York Times*, June 27, 1909, Magazine Section, 4.

201 **silk underwear, wool shirts and pants:** On plans for clothing and food, see *The World* (New York), July 4, 1909, Magazine Section, 2, and *Springfield Homestead* (Springfield, MA), March 27, 1909, 2.

201 **a training regimen:** *The Sun* (New York), May 25, 1909, 3.

201 **"We will be talking to the people":** *The San Francisco Examiner*, May 25, 1909, 2.

201 **"Berson and Süring":** A. Lawrence Rotch to David Todd, May 25, 1909, DPTP, Box 66, Folder 490.

201 **"Of course you don't believe":** David Todd to A. Lawrence Rotch, May 27, 1909, ALRP, Box 2.

201 **dismissed as "mere balloonacy":** *The Press* (Philadelphia), May 20, 1909, 6.

201 **"There is . . . every reason to believe":** *The New York Herald*, May 6, 1909, 9.

202 **"I suppose you and the Professor":** Lilian Whiting to Mabel Loomis Todd, May 12, 1909, MLTP, Box 24, Folder 734.

202 **Eager onlookers arrived by carriage and automobile:** Details of the day come from entry for August 11, 1909, diary of Mabel Loomis Todd, MLTP, Diaries and Journals Microfilm, Reel 5; entries for August 15, 16, 23, and September 2, 1909, journal of Mabel Loomis Todd, 67–89, MLTP, Diaries and Journals Microfilm, Reel 9; photographs, TBPC, Boxes 143–144, Folders 914–934; *The Boston Daily Globe*, August 12, 1909, 8; *Springfield Daily Republican* (Springfield, MA), August 12, 1909, 9; *Worcester Telegram*, August 12, 1909, 1 and 5; and *The Boston Post*, August 12, 1909, 2.

203 **"When you visit me, I want to talk":** W. W. Campbell to George Ellery Hale, May 11, 1908, GEHP, Box 9, Folder 6.

203 **"I fully share your opinion":** George Ellery Hale to W. W. Campbell, May 20, 1908, GEHP, Box 9, Folder 6.

CHAPTER 17: ENDGAME

204 **"Stars and stripes nailed to North Pole":** *The Chicago Daily Tribune*, September 7, 1909, 1.

204 **PEARY DISCOVERS THE NORTH POLE:** *The New York Times*, September 7, 1909, 1.

NOTES ON SOURCES

204 **if he claimed to have come from Mars:** *The New York Times*, September 7, 1909, 5.
205 **"I have always been a staunch admirer":** *Buffalo Evening Times*, September 7, 1909, 2.
205 **"Do not trouble about Cook's story":** *The New York Times*, September 11, 1909, 1.
205 **"Next to the wrangle over the discovery":** *Democrat and Chronicle* (Rochester, NY), September 19, 1909, 12.
206 **"those of a charlatan":** *Evening Bulletin* (Philadelphia), March 20, 1908, 3.
206 **"Censure can hardly be too severe":** *Science*, April 23, 1909, 661.
206 **A prominent British astronomer:** *The Daily Telegraph* (London), January 6, 1908, 3.
206 **decorated with moon vines and sunflowers:** *The Chicago Daily Tribune*, August 21, 1909, 3.
207 **"The public should not be deceived":** *The New York Thrice-a-Week World*, August 20, 1909, 5.
207 **"no one at Yerkes could see":** *The New York Times*, August 17, 1924, Section 8, 6.
207 **a leading American scientist replicated:** Newcomb (1907).
207 **"all success in an enterprise":** Edward C. Pickering to Percival Lowell, February 21, 1894, RHCOD, Box 10, Letterbook A12.
208 **"If any of us should go to a wealthy man":** Edward C. Pickering, "Reading of resolution, and abstracts of letters re Mars," in typed minutes of meeting at Yerkes Observatory, August 21, 1909, RAAS, Subgroup 2, Box 6, Folder 8.
208 **"As the public through the misrepresentation":** Astronomical and Astrophysical Society of America (1910:316).
208 **Todd knew beforehand that a tongue-lashing was in store:** Although the meeting at Yerkes Observatory did not occur until late August 1909, the society's secretary, William J. Hussey, wrote to all members on July 3, passing along a statement from Edward C. Pickering, the president: "Astronomy has been much injured, recently, by erroneous statements of our present knowledge of Mars. These misstatements have been widely circulated in Europe and America.... It is impossible to secure complete refutation, but a resolution based upon the following statement, especially if passed unanimously, would be widely published and distributed." Pickering then offered a draft statement much like the one eventually released to the press. Hussey concluded his letter: "The opinion of members is requested as to the desirability of issuing such a statement, officially, at the time of our next meeting. What would be your feeling in this matter?" David Todd wrote "yes" on his copy of the letter on July 10 and returned it to Hussey (RAAS, Subgroup 2, Box 4, Folder 9). Given that American astronomy was a small and intimate field, it is likely that Todd learned by word of mouth that a professional backlash was building against his Mars communications plan even well before he received the July 3 letter.
208 **"My observation of the 'canals'":** *Boston Evening Transcript*, May 22, 1909, Last Edition, Part Three, 2. Several weeks earlier, Todd told a reporter that he had come to doubt the artificiality of the lines on Mars because, if they were actual irrigation canals, they should not be straight: "Such canals would be constructed to follow the topography of the country. They would be contour lines.... Is it reasonable to suppose those canals, presenting straight, even and unbroken lines, are dug through mountains? If so what tremendous excavation there must be, and why do these unusual excavations not show through the telescope more vividly than the sections in the lower places? It is inconceivable and I think the theory must fall." *The Boston American*, May 2, 1909, 8S.
208 **"I have grave doubts if life":** *The New York Times*, June 27, 1909, Magazine Section, 4.
209 **blaming a lack of funds:** *Cleveland Plain Dealer*, October 9, 1909, 2; *The Hartford Courant*, October 20, 1909, 18.
209 **and life-giving oxygen too:** *The New York Times*, September 10, 1909, 14.

NOTES ON SOURCES

209 **"The mere attempt to use open canals":** Wallace (1907:105).
209 **"would probably be covered":** P. Lowell (1908:209).
210 **"To which of the two Americans":** *Le Petit Journal* (Paris), September 19, 1909, Supplément Illustré, 298.
210 **"The Eskimos laughed":** *The New York Times*, October 13, 1909, 1. See more at Herbert (1989:276–278).
210 **Eugène-Michel Antoniadi:** McKim (2018); McKim (1993a); McKim (1993b).
211 **"Had it not been for Prof. Schiaparelli's":** Antoniadi (1898b:63).
211 **calling himself "agnostic":** *English Mechanic*, August 28, 1903, 64.
211 **"Regarding the objectivity of the canals":** *The Journal of the British Astronomical Association* 18 (1907–1908): 401.
211 **the Meudon Observatory:** Dollfus (2013:7).
212 **larger than the full moon:** Sheehan and O'Meara (2001:167).
213 **"Its appearance was stunning":** E. M. Antoniadi to Giovanni Schiaparelli, September 21, 1909, FSDG, Cartella 1, Lettera 6.
213 **sensed that he must be dreaming:** Antoniadi (1916:32).
213 **"a succession of a few very faint":** Antoniadi (1916:84).
213 **"Geometry was conspicuous":** Antoniadi (1909b:79).
213 **"When I saw Mars at Meudon":** E. M. Antoniadi to E. E. Barnard, January 16, 1911, EEBP, Box 2.
213 **one of the top players in France:** *The British Chess Magazine*, September 1907, 413–414.
214 **sherry and cream peppermints:** Percival Lowell to H. Jevne, Nov. 5 and 19, 1909, LOAC, Copybook 10; 116 and 141.
214 **"I am extremely indebted to you":** E. M. Antoniadi to Percival Lowell, October 9, 1909, PLP, Percival Lowell Correspondence.
215 **"Dear Mr. Antoniadi":** Percival Lowell to E. M. Antoniadi, November 2, 1909, LOAC, Copybook 10, 108.
216 **The lenses were too big, he argued:** Hoyt (1976:48–49); P. Lowell (1894:539); Sheehan (1996:138–143). Other diehard canalists parroted this argument; see Waldemar Kaempffert in *The New York Times*, January 26, 1916, 10.
216 **"Of course, both of us":** E. M. Antoniadi to Percival Lowell, November 15, 1909, PLP, Percival Lowell Correspondence.
217 **"These first observations do not confirm":** Antoniadi (1909c:838).
217 **"The spider's webs of Mars are doomed":** E. M. Antoniadi to W. H. Wesley, October 30, 1909, RASL.
217 **"The geometrical network of Dr. Lowell":** E. M. Antoniadi to E. E. Barnard, December 11, 1909, EEBP, Box 2.
217 **"I am convinced that where":** E. M. Antoniadi to Giovanni Schiaparelli, November 30, 1909, FSDG, Cartella 1, Lettera 18.
217 **Greece, France, England, and the United States:** *New-York Tribune*, January 17, 1910, 6, and April 24, 1910, Part 5, 8; *Bulletin de la Société Astronomique de France*, November 1909, 488–494; *English Mechanic*, January 14, 1910, 561–562, and January 21, 1910, 584.
218 **staged photographs and faked diary entries:** Hartman (2023:273–277).
218 **Cook's evidence was "inexcusably lacking":** *The New York Times*, December 22, 1909, 1–2.
218 **"Maybe the doctor has gone":** *Illinois State Register* (Springfield), January 7, 1910, Weekly Edition, 8.
218 **Peary, too, was likely a cheat:** Hartman (2023:323).
218 **Sion College:** *The Illustrated London News*, October 9, 1886, 381.
219 **the agenda turned to Mars:** For details of the meeting, see *The Journal of the British Astronomical Association* 20 (1909–1910): 119–128.
219 **"The true appearance of the planet Mars":** Antoniadi (1909a:136).

220 **"We shall not expatiate on the circumstances":** Antoniadi (1909a:141).
220 **"There never was any real ground":** *Daily Express* (London), December 30, 1909, 1.
221 **"You may sleep quietly in your beds":** *The Washington Post*, January 16, 1910, 17; *The Vancouver Daily Province* (Vancouver, BC), January 31, 1910, 12; *Marlborough Express* (Blenheim, New Zealand), February 15, 1910, 6.
221 **"It doesn't interest me in the least":** *Boston Evening Transcript*, December 30, 1909, Last Edition, 12.
221 **"Mars has been imposing":** *Los Angeles Herald*, December 31, 1909, 4.

CHAPTER 18: POETIC ACHIEVEMENT

222 **"the crowning argument against the 'canals'":** *The San Francisco Examiner*, January 20, 1910, 18.
222 **"[They] teach us nothing at all":** G. Schiaparelli (1910). Translation from Crowe (1986:545).
223 **when the bright sun shone:** *The Astrophysical Journal*, November 1910, 319.
223 **remembrance of his great mentor:** P. Lowell (1910a).
223 **discovered two new canals:** P. Lowell (1910b); *Boston Evening Transcript*, December 31, 1909, Last Edition, 1–2.
223 **"It is not thinkable that":** *The Washington Post*, November 14, 1909, Magazine Section, 4.
223 **"Let's not rush to attribute":** *The New York Herald* (European Edition, Paris), March 23, 1910, 6.
223 **"The difficulty in establishing":** *University Missourian* (Columbia), February 18, 1910, 1.
224 **"In this recent art exhibition":** *The Outlook*, March 29, 1913, 719.
224 **"Personally I feel that so much":** Ray Long to Percival Lowell, August 3, 1910, PLP, Percival Lowell Correspondence.
224 **confessed to another,** *entre nous*: Ralph Wormeley Curtis to Barrett Wendell, February 2, 1910, WFP. Although Curtis wrote the letter in English, he switched to French when he commented on Lowell's doings, as if sharing a secret: "Entre nous Mars m'embête à la mort!" (Between us, Mars annoys me to death!)
224 **"what the French call 'Vers Libre'":** A. Lowell (1914:x).
224 **revisit their estimates for the age of the Earth:** Burchfield (1990).
224 **new breed of anthropologists:** C. King (2019).
225 **"Some of the consequences deducible":** *New York Evening Journal*, January 29, 1919, 20.
225 **"With fine constancy Percival Lowell":** *The Christian Science Monitor*, December 8, 1915, 24.
225 **"He leaps from point to point":** *The Sunday Herald* (Boston), May 17, 1914, Special Features Section, 6.
225 **"I have been down and up and down":** Percival Lowell to Amy Lowell, January 17, 1914, ALC.
225 **"It is nervous exhaustion":** W. L. Leonard to V. M. Slipher, December 11, 1912, VMSP, V. M. Slipher Correspondence.
225 **"The doctor says it is":** Constance Lowell to V. M. Slipher, October 27, 1912, VMSP, V. M. Slipher Correspondence.
225 **touchiness and tantrums:** "With Lowell came abuse, brutality, arrogance, conceit and insolence. It was my lot to work in the dome with him, and the night I was with him he knocked a light out of my hand, jerked a box out of my hand without asking me for it, pushed me backward across a pile of boxes and stepped over me before I could rise. Every order that he gave could have [been] heard a

half a mile away." W. H. Spaulding to A. E. Douglass, February 16, 1914, AEDP, Box 18, Folder 1 ("Lowelliana").
225 **"His health then was not very satisfactory":** *The New York Herald* (European Edition, Paris), November 15, 1916, 2.
225 **"The tendency to relapse":** Allbutt (1900:157).
226 **"Neurasthenia melts by infinite gradations":** Allbutt (1900:152).
226 **"We want war!":** *The Little Review*, October 1914, 7.
226 **"Boom! The Cathedral is a torch":** A. Lowell (1916:232).
226 **"In those two decades mankind":** *The New York Times*, September 19, 1916, 10.
226 **"present war of the world":** Percival Lowell to Hall Caine, November 19, 1914, LOAC, Copybook 14, 129.
226 **"Toll the bells in the steeples":** A. Lowell (1916:248–249).
227 **"Mars' Hill is a veritable colony":** Percival Lowell to F. J. Stimson, January 25, 1916, FJSL.
227 **"Humanity has not, thus far":** *The New York Times*, October 18, 1916, 10. Other newspapers expressed similar sentiments: "Do the Martians realize that a war is going on full swing on old planet Earth? Or, being the occupants of an older planet, have they progressed so far in intelligence and civilization that they have actually forgotten that there can be such a thing as war?" (*The Washington Herald*, January 10, 1915, Feature Section, 12.) "It is disheartening to hear a long-faced and sad-voiced scientist get up and say that the people on Mars have not long to live; but how much sadder must it be for some such person to arise on Mars and say that while we might live here happily till the end of time we are such damned fools that we have set about killing one another off and turning the planet back to the beasts of the field? Can you not hear Mars groan?" (*St. Louis Post-Dispatch*, February 6, 1916, Part Two, 1.)
227 **"The ear of the next generation":** Percival Lowell to Edward S. Morse, June 16, 1904, ESMP, E2, Box 9, Folder 1.
227 **Lowell looked even older:** Sheehan (2016).
227 **"Our observations have convinced us":** *The Daily Palo Alto* (Stanford University), October 18, 1916, 3.
228 **"This doesn't mean that the Martians":** *The Spokesman-Review* (Spokane), October 5, 1916, 6.
228 **"Great Discoveries and Their Reception":** *The Sunday Oregonian* (Portland), October 8, 1916, 21.
228 **"The undesired outsider is ignored":** Percival Lowell, "Great Discoveries and Their Reception," 13, PLP, Unpublished Lectures, Manuscripts, and Poems, Box 3U, Folder 11.
228 **"Gauge your work by its truth":** Percival Lowell, "Great Discoveries and Their Reception," 20–21, PLP, Unpublished Lectures, Manuscripts, and Poems, Box 3U, Folder 11.
228 **the Lick stargazers still ridiculed Lowell:** A month after dining with Lowell, Lick Observatory director W. W. Campbell derided the canal theory in a public lecture: "I doubt if there are any human hands on any portion of Mars. . . . The evidence in favor of such a theory is so meagre as to be practically non-exist[e]nt." *San Francisco Examiner*, November 11, 1916, 7.
228 **"Lick and Berkeley have now":** Percival Lowell to W. L. Leonard [October 17, 1916], in Leonard (1921:156). The dinner was at the home of Professor Armin Otto Leuschner, dean of the University of California Berkeley Graduate School, on October 16, 1916. See A. O. Leuschner to Percival Lowell, October 14, 1916, PLP, Percival Lowell Correspondence.
228 **news of his successful speaking tour:** *The Coconino Sun* (Flagstaff), October 20, 1916, 1.

228 **Pumpkins from his garden:** Percival Lowell to W. L. Leonard [circa late October 1916], in Leonard (1921:157).
228 **"Universities from Texas to Maine":** Percival Lowell to W. L. Leonard, November 11, 1916, in Leonard (1921:161).
229 **"Kind friends," the reverend began:** Father Cyprian Vabre, "Percival Lowell," PLP, Personal, Box 1, Folder 17. For more on Lowell's funeral, see *The Coconino Sun* (Flagstaff), November 24, 1916, 1.
229 **in his casket in academic regalia:** Benfey (2003:208).
229 **[AN ASTRONOMER] MUST LEARN TO WAIT:** The inscription comes from Lowell's magnum opus, *Mars and Its Canals*. P. Lowell (1906:10).
229 **"I should rather like to read":** Percival Lowell to Barrett Wendell, October 3 [1876], WFP.
229 **"He got neither the fun nor the glory":** Ralph Wormeley Curtis to Isabella Stewart Gardner, December 10 [1916], ISGP.
229 **"whether the explanation of":** *The Seattle Sunday Times*, December 17, 1916, Magazine Section, 4.
230 **"The human understanding":** F. Bacon (1861:56).
231 **"Whether Lowell has added anything":** *The Hartford Courant*, November 16, 1916, 8.
231 **"His work is a great and lasting one":** *Cleveland Plain Dealer*, November 17, 1916, 10.
231 **"Professor Lowell gave untrained men":** *The Paterson Press-Guardian* (Paterson, NJ), December 1, 1916, 6.
231 ***"Poeta nascitur, non fit":*** Constance Lowell, "Percival Lowell, gleaned by his wife," PLP, Biographies and Bibliography, Box 1.
231 **"The Stars have claimed you":** Irva Struthers McCall, "In memoriam: P.L. 1916," PLP, Personal, Box 3.

EPILOGUE: CHILDREN OF MARS

232 **"I am expecting to [write] the Institute of France":** *The New York Times*, July 11, 1937, Section 2, 2.
232 **pressed him into early retirement:** Dobrow (2020:240).
232 **suffering from "circular insanity":** Entry for September 14, 1922, journal of Mabel Loomis Todd, MLTP, Diaries and Journals Microfilm, Reel 9. Mabel Todd reported further: "Last spring in Cocoanut Grove [Florida] my dear David grew queerer and queerer. . . . He brought the strangest people to the house, talked incessantly, and finally went to Nassau [Bahamas] at five minutes' notice, and stayed ten days. . . . He soon gave up going to bed and eating at all."
233 **"WHY DO YOU KEEP ME locked up":** David Todd to Mabel Loomis Todd, August 13, 1922, MLTP, Box 37, Folder 1086.
233 **"I can't write of it":** Entry for September 14, 1922, journal of Mabel Loomis Todd, MLTP, Diaries and Journals Microfilm, Reel 9.
233 **high-flying balloons:** In 1919, David Todd resurrected plans for a high-altitude ascension with aeronaut Leo Stevens. *The New York Times*, July 8, 1919, 5; *The Omaha Sunday Bee*, October 19, 1919, Part Three, 1; *The Sun* (New York), October 26, 1919, Magazine Section, 3. In 1924, he requested—but was denied—the use of a U.S. Navy dirigible for the experiment. *The Philadelphia Inquirer*, August 22, 1924, 7.
233 **an underground telescope:** *Boston Sunday Post*, August 28, 1921, 29.
233 **the U.S. military to silence its radios:** *The New York Times*, August 21, 1924, 11; *The Washington Post*, August 22, 1924, 3; *The Christian Science Monitor*, August 23, 1924, 1.

233 **None were heard:** *The Washington Post*, August 29, 1924, 13. Although the military detected no Martian signals, David Todd picked up strange transmissions that he believed proved "beyond a shadow of a doubt, that intelligence ... exists on the planet Mars." Godfrey (2014:115).
233 **"Our world is radically imperfect":** *L'Astronomie*, May 1917, 167.
233 **"Rest assured, we shall some day":** *The New York Times*, December 12, 1923, 3.
233 **"a new, a more glorious, resurrection":** *Springfield Weekly Republican* (Springfield, MA), May 30, 1940, 3.
233 **she still found Lowell's vision of the planet to be credible:** On what would have been Lowell's eighty-fifth birthday, Whiting wrote, "The charm of Dr. Lowell's repeated lectures before the Lowell institute in Boston can never fade from the mind of any attendant; his theory—not yet either proved or denied—of Mars ... held an audience almost breathless." *Springfield Daily Republican* (Springfield, MA), March 13, 1940, 8.
233 **"Perhaps it is to some such place":** *Boston Sunday Post*, December 11, 1910, 26.
233 **"Some of you may have read a book":** Wells (1937:62).
234 **the premise struck some in the audience as all too real:** Schwartz (2015).
234 **"Has not science proved":** Gerta S. Brown to Mercury Theater, November 2, 1938 [date received], RWOWP, Box 24, New York (State) Folder 2.
234 **"Now the story of the 'canals'":** Waterfield (1938:50).
234 **"This controversy brought disrepute":** Condon (1969:841).
234 **"Are there people on Mars?":** *New York Evening Journal*, March 2, 1922, 24.
235 **"I imagined how wonderful":** Goddard and Pendray (1970:9). In 1932, Goddard wrote a fan letter to H. G. Wells: "In 1898 I read your *War of the Worlds*. I was sixteen years old, and [it] made a deep impression. The spell was complete about a year afterward, and I decided that what might conservatively be called 'high-altitude research' was the most fascinating problem in existence. The spell did not break, and I took up physics as a profession." Goddard and Pendray (1970:821–823).
235 **"lapsed into delirium, raving":** Moskowitz (1963:229), who writes that Gernsback was inspired by Lowell's *Mars as the Abode of Life*, which is impossible because the book was not published until 1908 (as noted by Westfahl [1998:137–138]). The book that Gernsback actually read, according to Gernsback himself, was Lowell's 1895 work, *Mars*. See *Life*, July 26, 1963, 64.
235 **a genre he called "scientifiction":** Moskowitz (1963:313–322).
236 **"I was about twelve then":** Richardson (1954:154–155).
236 **"I looked at that magazine":** *Science Fiction Review*, May 1977, 7.
236 **"not only possible, but even probable":** *Pawtuxet Valley Gleaner* (Warwick, RI), September 7, 1906, 1. Lovecraft later insisted that he had never believed Lowell's speculations about Mars. See Joshi (2010:121–123).
236 **Ray Bradbury, Arthur C. Clarke, and Leigh Brackett:** Seed (2015:45–47); Bradbury et al. (1973:17–18, 27); Newell and Lamont (2011).
237 **"I can remember spending":** Sagan (1980:111).
237 **"The fact is, that ideas":** P. Lowell (1895c:318–319).
237 **"Whatever we can say about Lowell's":** Bradbury et al. (1973:27).
238 **"I think it's part of the nature of man":** Bradbury et al. (1973:35).
238 **"I don't think there is an advanced civilization":** Bradbury et al. (1973:32–33).
238 **mere sand paintings:** *Popular Science*, September 1972, 51–55.
238 **correlated with nothing on the Martian surface:** Sagan and Fox (1975).
238 **"Even if all Lowell's conclusions":** Sagan (1980:111).
239 **geological processes have largely halted:** S. S. Johnson (2020:168).
239 **most of its atmosphere:** Jakosky (2019).
239 **Mars may have been warm and earthlike:** Ramirez and Craddock (2018).
239 **a meteorite in Antarctica:** McKay et al. (1996).

239 **modern-day Mars "boom":** *The New York Times*, August 7, 1996, 1 and 10; *The Washington Post*, August 8, 1996, A3.
240 **we may all be of Martian birth:** Loeb (2021:168–172).
240 **reengineer the entire globe:** *The New York Times*, January 4, 2022, D5.
240 **for less than the cost of a middle-class home:** Musk (2017).
240 **"The time has come":** Zubrin (2024:280).
241 **"laid out in a spider-web shape":** Crossman (2021:44).
241 **"based on solidarity, benevolence":** Crossman (2021:199).

Select Bibliography

Allbutt, Thomas Clifford, ed. 1900. *A System of Medicine by Many Writers.* Vol. 8. New York: Macmillan.
Anderson, Lawrence B. 1984. "The Rogers Building: 1866–1938." *Places* 1 (4):38–46.
Antoniadi, E. M. 1897. "On the Rotation Period of Venus." *The Journal of the British Astronomical Association* 8 (1):43–46.
Antoniadi, E. M. 1898a. "Notes on the Rotation Period of Venus." *Monthly Notices of the Royal Astronomical Society* 58 (5):313–320.
Antoniadi, E. M. 1898b. "Section for the Observation of Mars. Report of the Section, 1896." *Memoirs of the British Astronomical Association, Reports of the Observing Sections* 6:55–102.
Antoniadi, E. M. 1902. "Recent Observations of Mars." *Knowledge* 25 (198):81–84.
Antoniadi, E. M. 1909a. "Mars Section: Fifth Interim Report for 1909, Dealing with the Fact Revealed by Observation that Prof. Schiaparelli's 'Canal' Network is the Optical Product of the Irregular Minor Details Diversifying the Martian Surface." *The Journal of the British Astronomical Association* 20 (3):136–141.
Antoniadi, E. M. 1909b. "Mars Section: Fourth Interim Report for the Apparition of 1909, Dealing with the Appearance of the Planet Mars Between September 20 and October 23 in the Great Refractor of the Meudon Observatory." *The Journal of the British Astronomical Association* 20 (2):78–81.
Antoniadi, E. M. 1909c. "Observations de la Planète Mars Faites à l'Observatoire de Meudon." *Comptes Rendus des Séances de l'Académie des Sciences* 149:836–838.
Antoniadi, E. M. 1916. "Section for the Observation of Mars. Report of the Section, 1909." *Memoirs of the British Astronomical Association* 20:25–92.
Antoniadi, E. M. 1975. *The Planet Mars*, translated by Patrick Moore. Shaldon, Devon, U.K.: Keith Reid.
Arnold, Edwin. 1894. "Astronomy and Religion." *The North American Review* 159 (455):404–415.
Astronomical and Astrophysical Society of America. 1910. *Publications of the Astronomical and Astrophysical Society of America.* Vol. 1: *Organization, Membership, and Abstracts of Papers 1897–1909.* Published by the society.
Atchison, Topeka, and Santa Fe Railway Company. 1898. *A Colorado Summer.* Chicago: Corbitt and Butterfield.
Bacon, Francis. 1861. *The Philosophical Works of Francis Bacon.* Vol. 4, edited by James Spedding. London: Longman.
Bacon, Gertrude. 1898. "A Girl's Balloon Journey over London." *The Harmsworth Magazine* 1 (4):400–404.

Baedeker, Karl. 1908. *London and Its Environs*. Leipzig: Karl Baedeker.
Bailey, Nettie. 1905. "The Significance of the Woman's Club Movement." *Harper's Bazar* 39 (3):204–209.
Baker, Ray Stannard. 1902. "Marconi's Achievement. Telegraphing Across the Ocean Without Wires." *McClure's Magazine* 18 (4):291–299.
Ball, Robert S. 1894. "The Possibility of Life in Other Worlds." *The Fortnightly Review* [New Series] 56 (335):718–729.
Ball, Robert S. 1915. *Reminiscences and Letters of Sir Robert Ball*, edited by W. Valentine Ball. London: Cassell and Co.
Barber, Richard. 1973. *King Arthur: In Legend and History*. Ipswich: Boydell Press.
Barrett, James Wyman. 1941. *Joseph Pulitzer and His World*. New York: Vanguard.
Basalla, George. 2006. *Civilized Life in the Universe: Scientists on Intelligent Extraterrestrials*. New York: Oxford University Press.
Bassan, Maurice. 1970. *Hawthorne's Son: The Life and Literary Career of Julian Hawthorne*. Columbus: Ohio State University Press.
Beck, Peter J. 2016. *The War of the Worlds: From H. G. Wells to Orson Welles, Jeff Wayne, Steven Spielberg and Beyond*. London: Bloomsbury.
Bell, Alexander Graham. 1910. "The Neighboring Worlds." *The Volta Review* 12 (9):549–552.
Benfey, Christopher. 2003. *The Great Wave: Gilded Age Misfits, Japanese Eccentrics, and the Opening of Old Japan*. New York: Random House.
Benstock, Shari. 1994. *No Gifts from Chance: A Biography of Edith Wharton*. Austin: University of Texas Press.
Berger, Meyer. 1951. *The Story of the New York Times, 1851–1951*. New York: Simon & Schuster.
Bernardi, Gabriella. 2017. *Giovanni Domenico Cassini: A Modern Astronomer in the 17th Century*. Cham, Switzerland: Springer.
Blakemore, Harold. 1974. *British Nitrates and Chilean Politics, 1886–1896: Balmaceda and North*. London: Athlone Press.
Blum, Deborah. 2006. *Ghost Hunters: William James and the Search for Scientific Proof of Life After Death*. New York: Penguin.
Blumenthal, Walter Hart. 1932. "Books Bound in Human Skin." *The American Book Collector* 2:119–124.
Blumenthal, Walter Hart. 1955. *Bookmen's Bedlam: An Olio of Literary Oddities*. New Brunswick: Rutgers University Press.
Boston Public Library. 1889. *City of Boston. Thirty-Seventh Annual Report of the Trustees of the Public Library*. Boston: Boston Public Library.
Bowler, Peter J. 2012. "Christian Responses to Darwinism in the Late Nineteenth Century." In *The Blackwell Companion to Science and Christianity*, edited by J. B. Stump and Alan G. Padgett, 37–47. Chichester, U.K.: Wiley-Blackwell.
Bradbury, Ray. 1950. *The Martian Chronicles*. New York: Doubleday.
Bradbury, Ray, et al. 1973. *Mars and the Mind of Man*. New York: Harper and Row.
Breitbart, Eric. 1997. *A World on Display: Photographs from the St. Louis World's Fair, 1904*. Albuquerque: University of New Mexico Press.
[Brisbane, Arthur]. 1906. *Editorials from the Hearst Newspapers*. New York: Albertson Publishing Co.
Bruner, Robert F., and Sean D. Carr. 2023. *The Panic of 1907: Heralding a New Era of Finance, Capitalism, and Democracy*. Second Edition. Hoboken: Wiley.
Burchfield, Joe D. 1990. *Lord Kelvin and the Age of the Earth*. Chicago: University of Chicago Press.
Burroughs, Edgar Rice. 1917. *A Princess of Mars*. New York: Grosset and Dunlap.

Campbell, W. Joseph. 2001. *Yellow Journalism: Puncturing the Myths, Defining the Legacies.* Westport, CT: Praeger.

Campbell, W. W. 1894. "An Explanation of the Bright Projections Observed on the Terminator of Mars." *Publications of the Astronomical Society of the Pacific* 6 (35):103–112.

Cantor, Geoffrey. 2006. "Anglo-Jewish Responses to Evolution." In *Jewish Tradition and the Challenge of Darwinism*, edited by Geoffrey Cantor and Marc Swetlitz, 23–46. Chicago: University of Chicago Press.

Carlson, Oliver. 1937. *Brisbane: A Candid Biography.* New York: Stackpole Sons.

Carlson, W. Bernard. 2013. *Tesla: Inventor of the Electrical Age.* Princeton: Princeton University Press.

Carpino, Mario. 2017. "The Merz Refractors at the Brera Astronomical Observatory." In *Merz Telescopes: A Global Heritage Worth Preserving*, edited by Ileana Chinnici, 87–100. Cham, Switzerland: Springer.

Cassini, Giovanni Domenico. 1666. *Martis Circa Axem Proprium Revolubilis Observationes Bononiae.* Bologna: HH. de Ducciis.

Chapman, Allan. 2007. "Sir Robert Stawell Ball (1840–1913): Royal Astronomer in Ireland and Astronomy's Public Voice." *Journal of Astronomical History and Heritage* 10 (3):198–210.

Cheney, Margaret. 1981. *Tesla: Man Out of Time.* Englewood Cliffs, NJ: Prentice-Hall.

Cheney, Margaret, and Robert Uth. 1999. *Tesla, Master of Lightning.* New York: Barnes and Noble Books.

Churchill, Allen. 1958. *Park Row.* New York: Rinehart and Co.

Clarke, Arthur C. 1967. *The Sands of Mars.* New York: Harcourt Brace Jovanovich.

Clary, David A. 2003. *Rocket Man: Robert H. Goddard and the Birth of the Space Age.* New York: Theia.

Cline, Platt. 1976. *They Came to the Mountain: The Story of Flagstaff's Beginnings.* Flagstaff: Northern Arizona University.

Cline, Platt. 1994. *Mountain Town: Flagstaff's First Century.* Flagstaff: Northland.

Condon, Edward U. 1969. *Final Report of the Scientific Study of Unidentified Flying Objects Conducted by the University of Colorado Under Contract to the United States Air Force.* New York: E. P. Dutton and Co.

Cooper, Christopher. 2015. *The Truth About Tesla: The Myth of the Lone Genius in the History of Innovation.* New York: Race Point.

Cotardière, Philippe de la, and Patrick Fuentes. 1994. *Camille Flammarion.* Paris: Flammarion.

Cowper, H. S. 1897. *The Hill of the Graces: A Record of Investigation Among the Trilithons and Megalithic Sites of Tripoli.* London: Methuen and Co.

Croll, James. 1890. *The Philosophical Basis of Evolution.* London: Edward Stanford.

Crossley, Robert. 2011. *Imagining Mars: A Literary History.* Middletown, CT: Wesleyan University Press.

Crossman, Frank, ed. 2021. *Mars City States: New Societies for a New World.* Lakewood, CO: Polaris.

Crowe, Michael J. 1986. *The Extraterrestrial Life Debate, 1750–1900.* Cambridge, U.K.: Cambridge University Press.

Cuthbert, Guillermo Burgos. 2011. *Oficina Alianza: Memoria Fotográfica de una Salitrera de Tarapacá.* Santiago: Ricaaventura.

Damon, S. Foster. 1935. *Amy Lowell: A Chronicle with Extracts from Her Correspondence.* Boston: Houghton Mifflin.

Davenhall, Clive. 2018. "Mars and the Mediums." In *Imagining Other Worlds: Explorations* in *Astronomy and Culture*, edited by Nicholas Campion and Chris Impey, 107–118. Ceredigion, Wales: Sophia Centre Press.

Davidson, Edward H., and William J. Scheick. 1994. *Paine, Scripture, and Authority:* The Age of Reason *as Religious and Political Idea*. Bethlehem: Lehigh University Press.

Davidson, Keay. 1999. *Carl Sagan: A Life*. New York: John Wiley and Sons.

DeVorkin, David H., ed. 1999. *The American Astronomical Society's First Century*. Washington, D.C.: American Institute of Physics.

Dick, Steven J. 1982. *Plurality of Worlds: The Origins of the Extraterrestrial Life Debate from Democritus to Kant*. Cambridge, U.K.: Cambridge University Press.

Dick, Steven J. 1996. *The Biological Universe: The Twentieth-Century Extraterrestrial Life Debate and the Limits of Science*. Cambridge, U.K.: Cambridge University Press.

Dick, Steven J. 2012. "Space, Time and Aliens: The Role of Imagination in Outer Space." In *Imagining Outer Space: European Astroculture in the Twentieth Century*, edited by Alexander C. T. Geppert, 27–44. New York: Palgrave Macmillan.

Dick, Steven J. 2018. *Astrobiology, Discovery, and Societal Impact*. Cambridge, U.K.: Cambridge University Press.

Dobrow, Julie. 2017. "The Star-Crossed Astronomer." *Amherst* 69 (4):22–27.

Dobrow, Julie. 2018. *After Emily: Two Remarkable Women and the Legacy of America's Greatest Poet*. New York: W. W. Norton.

Dobrow, Julie. 2020. "Eclipses, Ecology, and Emily Dickinson." In *Amherst in the World*, edited by Martha Saxton, 235–248. Amherst, MA: Amherst College Press.

Doctorow, E. L. 1975. *Ragtime*. New York: Random House.

Dollfus, Audouin. 2013. *The Great Refractor of Meudon Observatory*, translated by Richard McKim. New York: Springer Science+Business Media.

Douglass, Andrew Ellicott. 1907a. "Illusions of Vision and the Canals of Mars." *The Popular Science Monthly* 70 (5):464–474.

Douglass, Andrew Ellicott. 1907b. "Is Mars Inhabited?" *The Harvard Illustrated Magazine* 8 (6):116–118.

Dumas, F. 1888. *Le Commandant Guzman*. Paris: E. Plon, Nourrit, and Co.

Evans, J. E., and E. Walter Maunder. 1903. "Experiments as to the Actuality of the 'Canals' Observed on Mars." *Monthly Notices of the Royal Astronomical Society* 63 (8):488–499.

Everett, Edward. 1840. *A Memoir of John Lowell, Jr*. Boston: Charles C. Little and James Brown.

Fine, Barnett. 1933. *A Giant of the Press*. New York: Editor and Publisher Library.

Flammarion, Camille. 1862. *La Pluralité des Mondes Habités*. Paris: Mallet-Bachelier.

Flammarion, Camille. 1877. *Les Terres du Ciel*. Paris: C. Marpon et E. Flammarion.

Flammarion, Camille. 1891. *Urania*, translated by Augusta Rice Stetson. London: Chatto and Windus.

Flammarion, Camille. 1892a. "Inter-Astral Communication." *The New Review* 6 (32):106–114.

Flammarion, Camille. 1892b. *La Planète Mars et Ses Conditions d'Habitabilité*. Paris: Gauthier-Villars et Fils.

Flammarion, Camille. 1894. *Popular Astronomy: A General Description of the Heavens*, translated by J. Ellard Gore. New York: D. Appleton and Co.

Flammarion, Camille. 1897. "Is Mars Inhabited?" *The Humanitarian* 1 (5):309–317.

Flammarion, Camille. 1907. *Mysterious Psychic Forces: An Account of the Author's Investigations in Psychical Research, Together with Those of Other European Savants*. Boston: Small, Maynard and Co.

Flammarion, Camille. 1911. *Mémoires Biographique et Philosophiques d'un Astronome*. Paris: Flammarion.

Flammarion, Camille. 2015. *Camille Flammarion's The Planet Mars. As Translated by Patrick Moore*. Cham, Switzerland: Springer.

Flammarion, Mme. Camille [Sylvie]. 1893. "Étrange Surprise." *L'Initiation* 18 (5):137–144.

Flournoy, Théodore. 1900. *From India to the Planet Mars: A Study of a Case of Somnambulism*, translated by Daniel B. Vermilye. New York: Harper and Brothers.

Flournoy, Théodore. 1994. *From India to the Planet Mars: A Case of Multiple Personality with Imaginary Languages*, with a foreword by C. G. Jung and commentary by Mireille Cifali. Princeton: Princeton University Press.

Fox, Stephen. 1997. *The Mirror Makers: A History of American Advertising and Its Creators*. Urbana: University of Illinois Press.

Franklin, Benjamin. 1961. *The Papers of Benjamin Franklin*. Vol. 3: *January 1, 1745, Through June 30, 1750*, edited by Leonard W. Labaree. New Haven: Yale University Press.

Fuller, Loie. 1913. *Fifteen Years of a Dancer's Life: With Some Account of Her Distinguished Friends*. Boston: Small, Maynard, and Co.

Ganthony, Richard. 1923. *A Message from Mars: A Fantastic Comedy in Three Acts*. New York: Samuel French.

Garczynski, Edward R. 1891. *Auditorium*. New York: Exhibit Publishing Co.

Gilman, Charlotte Perkins. 1905. *Women and Economics: A Study of the Economic Relation Between Men and Women as a Factor in Social Evolution*. Fourth Edition. Boston: Small, Maynard and Co.

Glaisher, James, et al. 1871. *Travels in the Air*. London: Richard Bentley.

Goddard, Esther C., and G. Edward Pendray, eds. 1970. *The Papers of Robert H. Goddard*. Vol. 1–3. New York: McGraw-Hill.

Godfrey, Donald G. 2014. *C. Francis Jenkins, Pioneer of Film and Television*. Urbana: University of Illinois Press.

Goodman, Susan. 2011. *Republic of Words: The Atlantic Monthly and Its Writers*. Hanover, NH: University Press of New England.

Gould, Jean. 1975. *Amy: The World of Amy Lowell and the Imagist Movement*. New York: Dodd, Mead and Co.

Greenslet, Ferris. 1946. *The Lowells and Their Seven Worlds*. Boston: Houghton Mifflin.

Gregory, R. A. 1900. "Mars as a World." *The National Review* 34 (204):914–922.

Griffith, M. 1896. "The Christopher Columbus of Mars." *Pearson's Magazine* 2 (7):30–37.

Grinspoon, David. 2004. *Lonely Planets: The Natural Philosophy of Alien Life*. New York: Ecco.

Haller, John S., Jr. 1971. *Outcasts from Evolution: Scientific Attitudes of Racial Inferiority, 1859–1900*. Urbana: University of Illinois Press.

Harris, Marvin. 2001. *The Rise of Anthropological Theory: A History of Theories of Culture*. Updated Edition. Walnut Creek, CA: AltaMira Press.

Hartman, Darrell. 2023. *Battle of Ink and Ice: A Sensational Story of News Barons, North Pole Explorers, and the Making of Modern Media*. New York: Viking.

Hawkins, Laurence A. 1903. "Nikola Tesla, His Work and Unfulfilled Promises." *The Electrical Age* 30 (2):99–108.

Hawley, Samuel, ed. 2008. *America's Man in Korea: The Private Letters of George C. Foulk, 1884–1887*. Lanham, MD: Lexington Books.

Henderson, Bruce. 2005. *True North: Peary, Cook, and the Race to the Pole*. New York: W. W. Norton.

Herbert, Wally. 1989. *The Noose of Laurels: Robert E. Peary and the Race to the North Pole.* New York: Atheneum.

Herschel, William. 1795. "On the Nature and Construction of the Sun and Fixed Stars." *Universal Magazine of Knowledge and Pleasure* 97:99–105.

Hetherington, Norriss S. 1976. "Amateur Versus Professional: The British Astronomical Association and the Controversy over Canals on Mars." *Journal of the British Astronomical Association* 86 (4):303–308.

Hildebrandt, A. 1908. *Airships Past and Present: Together with Chapters on the Use of Balloons in Connection with Meteorology, Photography and the Carrier Pigeon*, translated by W. H. Story. London: Archibald Constable and Co.

Holden, Edward S. 1892. "Note on the Mount Hamilton Observation of Mars, June–August, 1892." *Astronomy and Astro-Physics* 11 (8):663–668.

Holden, Edward S. 1894a. "Bright Projections at the Terminator of Mars." *Publications of the Astronomical Society of the Pacific* 6 (38):285–286.

Holden, Edward S. 1894b. "The Lowell Observatory, in Arizona." *Publications of the Astronomical Society of the Pacific* 6 (36):160–169.

Holden, Edward S. 1897. "Mr. Lowell's Observations of Mercury and Venus." *Science* [New Series] 5 (121):656–657.

Holden, Edward S. 1901. "What We Know About Mars." *McClure's Magazine* 16 (5):439–444.

Hollis, J. M. 1992. "Wrexie." *Griffith Observer* 56 (1):10–17.

Holmes, Oliver Wendell. 1861. *Elsie Venner: A Romance of Destiny.* Boston: Ticknor and Fields.

Holmes, Richard. 2013. *Falling Upwards: How We Took to the Air.* New York: Pantheon.

Holtz, Allan. 2012. *American Newspaper Comics: An Encyclopedic Reference Guide.* Ann Arbor: University of Michigan Press.

Hoyt, William Graves. 1976. *Lowell and Mars.* Tucson: University of Arizona Press.

Hughes, David Y., and Robert M. Philmus. 1973. "The Early Science Journalism of H. G. Wells: A Chronological Survey." *Science Fiction Studies* 1 (2):98–114.

Hunt, Inez, and Wanetta W. Draper. 1964. *Lightning in His Hand: The Life Story of Nikola Tesla.* Denver: Sage Books.

Hyslop, James H. 1913. "Journeys to the Planet Mars." *Journal of the American Society for Psychical Research* 7 (5):272–283.

Ings, Simon. 2007. *A Natural History of Seeing: The Art and Science of Vision.* New York: W. W. Norton.

International Geographical Congress. 1905. *Report of the Eighth International Geographic Congress Held in the United States 1904.* Washington, D.C.: Government Printing Office.

Iqbal, Muzaffar. 2009. "Darwin's Shadow: Context and Reception in the Muslim World." *Islam and Science* 7 (1):9–50.

Jakosky, Bruce. 1998. *The Search for Life on Other Planets.* Cambridge, U.K.: Cambridge University Press.

Jakosky, Bruce. 2006. *Science, Society, and the Search for Life in the Universe.* Tucson: University of Arizona Press.

Jakosky, Bruce. 2019. "The CO_2 Inventory on Mars." *Planetary and Space Science* 175:52–59.

James, William. 2001. *The Correspondence of William James.* Vol. 9: *July 1899–June 1901*, edited by Ignas K. Skrupskelis and Elizabeth M. Berkeley. Charlottesville: University Press of Virginia.

John, Arthur. 1981. *The Best Years of the Century: Richard Watson Gilder, Scribner's Monthly, and the Century Magazine, 1870–1909.* Urbana: University of Illinois Press.

Johnson, Robert Underwood. 1923. *Remembered Yesterdays.* Boston: Little, Brown and Co.

Johnson, Sarah Stewart. 2020. *The Sirens of Mars: Searching for Life on Another World.* New York: Crown.

Jones, Bessie Zaban, and Lyle Gifford Boyd. 1971. *The Harvard College Observatory.* Cambridge, MA: Belknap.

Jones, Roger. 2005. "Sir Robert Ball: Victorian Astronomer and Lecturer Par Excellence." *The Antiquarian Astronomer* 2:27–36.

Jones, Ruth Owen. 1994. "Those Daring Young Amherst Men (and One Intrepid Woman) in Their Flying Machines." *Amherst* 46 (4):28–31.

Joshi, S. T. 2010. *I Am Providence: The Life and Times of H. P. Lovecraft.* Vol. 1. New York: Hippocampus.

Juergens, George. 1966. *Joseph Pulitzer and the* New York World. Princeton: Princeton University Press.

Kaempffert, Waldemar. 1907. "What We Know About Mars." *McClure's Magazine* 28 (5):481–486.

Kepler, Johannes. 1965. *Kepler's Conversation with Galileo's Sidereal Messenger,* translated by Edward Rosen. New York: Johnson Reprint Corp.

Kinder, Anthony J. 2008. "Edward Walter Maunder FRAS (1851–1928): His Life and Times." *Journal of the British Astronomical Association* 118 (1):21–42.

King, Charles. 2019. *Gods of the Upper Air: How a Circle of Renegade Anthropologists Reinvented Race, Sex, and Gender in the Twentieth Century.* New York: Doubleday.

King, Greg. 2009. *The Season of Splendor: The Court of Mrs. Astor in Gilded Age New York.* Hoboken: John Wiley and Sons.

Kirkland, Joseph, and Caroline Kirkland. 1894. *The Story of Chicago.* Vol. 2. Chicago: Dibble Publishing Co.

Koizumi, Kazuo, ed. 1936. *Letters from Basil Hall Chamberlain to Lafcadio Hearn.* Tokyo: The Hokuseido Press.

Koizumi, Kazuo, ed. 1937. *More Letters from Basil Hall Chamberlain to Lafcadio Hearn, and Letters from M. Toyama, Y. Tsubouchi and Others.* Tokyo: The Hokuseido Press.

Labitzke, Karin G., and Harry van Loon. 1999. *The Stratosphere: Phenomena, History, and Relevance.* Berlin: Springer.

Lafertte, Elías. 2014. "Nitrate Workers and State Violence: The Massacre at Escuela Santa María de Iquique." In *The Chile Reader: History, Culture, Politics,* edited by Elizabeth Quay Hutchison, Thomas Miller Klubock, Nara B. Milanich, and Peter Winn, 238–244. Durham: Duke University Press.

Lampland, C. O. 1905. "On Photographing the Canals of Mars." *Lowell Observatory Bulletin No. 21*:136–137.

Lane, K. Maria D. 2011. *Geographies of Mars: Seeing and Knowing the Red Planet.* Chicago: University of Chicago Press.

Larson, Edward J. 2018. *To the Edges of the Earth: 1909, the Race for the Three Poles, and the Climax of the Age of Exploration.* New York: William Morrow.

Lears, T. J. Jackson. 1981. *No Place of Grace: Antimodernism and the Transformation of American Culture 1880–1920.* New York: Pantheon.

Leonard, Louise. 1921. *Percival Lowell: An Afterglow.* Boston: Richard G. Badger.

Lepper, George Henry. 1919. *From Nebula to Nebula; or, The Dynamics of the Heavens.* Fourth Edition. Pittsburgh: Privately Printed.

Lightman, Bernard. 2007. *Victorian Popularizers of Science: Designing Nature for New Audiences*. Chicago: University of Chicago Press.
Loeb, Avi. 2021. *Extraterrestrial: The First Sign of Intelligent Life Beyond Earth*. Boston: Houghton Mifflin Harcourt.
Longsworth, Polly. 1984. *Austin and Mabel: The Amherst Affair & Love Letters of Austin Dickinson and Mabel Loomis Todd*. New York: Farrar, Straus and Giroux.
Lowell, A. Lawrence. 1935. *Biography of Percival Lowell*. New York: Macmillan.
Lowell, Amy. 1914. *Sword Blades and Poppy Seed*. New York: Macmillan.
Lowell, Amy. 1916. *Men, Women and Ghosts*. New York: Macmillan.
Lowell, Percival. 1886. *Chosön, the Land of the Morning Calm: A Sketch of Korea*. Boston: Ticknor and Co.
Lowell, Percival. 1888. *The Soul of the Far East*. Boston: Houghton, Mifflin and Co.
Lowell, Percival. 1891. *Noto: An Unexplored Corner of Japan*. Cambridge, MA: Riverside Press.
Lowell, Percival. 1894. "Mars." *Astronomy and Astro-Physics* 13 (7):538–553.
Lowell, Percival. 1895a. *Mars*. Boston: Houghton, Mifflin and Co.
Lowell, Percival. 1895b. "Mars: The Flagstaff Photographs." *The New England Magazine* [New Series] 12 (6):643–655.
Lowell, Percival. 1895c. *Occult Japan, or The Way of the Gods: An Esoteric Study of Japanese Personality and Possession*. Boston: Houghton, Mifflin and Co.
Lowell, Percival. 1898. *Annals of the Lowell Observatory*. Vol. 1: *Observations of the Planet Mars During the Opposition of 1894–5*. Boston: Houghton, Mifflin and Co.
Lowell, Percival. 1902. "The Markings on Venus." *Astronomische Nachrichten* 160 (3823):129–132.
Lowell, Percival. 1903. *The Solar System: Six Lectures Delivered at the Massachusetts Institute of Technology in December, 1902*. Boston: Houghton, Mifflin and Co.
Lowell, Percival. 1904a. "The Cartouches of the Canals of Mars." *Lowell Observatory Bulletin No. 12*:59–86.
Lowell, Percival. 1904b. "The Thoth and the Amenthes." *Lowell Observatory Bulletin No. 8*:39–43.
Lowell, Percival. 1905a. "Double Canals of Mars in 1903." *Lowell Observatory Bulletin No. 15*:97–110.
Lowell, Percival. 1905b. "The Canals of Mars—Photographed." *Lowell Observatory Bulletin No. 21*:134–136.
Lowell, Percival. 1906. *Mars and Its Canals*. New York: Macmillan.
Lowell, Percival. 1907. "New Photographs of Mars." *The Century Magazine* 75 (2):303–311.
Lowell, Percival. 1908. *Mars as the Abode of Life*. New York: Macmillan.
Lowell, Percival. 1909. "The Revelation of Evolution. A Thought and Its Thinkers." *Atlantic Monthly* 104 (2):174–183.
Lowell, Percival. 1910a. "Schiaparelli." *Popular Astronomy* 18 (8):456–467.
Lowell, Percival. 1910b. "The Canali Novae of Mars." *Lowell Observatory Bulletin No. 45*:243–246.
Lucas, Charles J. P. 1905. *The Olympic Games, 1904*. St. Louis: Woodward and Tiernan Printing Co.
Mackay, Charles. 1841. *Memoirs of Extraordinary Popular Delusions*. Vol. 1. London: Richard Bentley.
Macpherson, Hector, Jr. 1905. *Astronomers of To-Day and Their Work*. London: Gall and Inglis.
Magri, Countess M. Lavinia. 1979. *The Autobiography of Mrs. Tom Thumb (Some of My Life Experiences)*, edited and introduced by A. H. Saxon. Hamden, CT: Archon Books.

Marion, Fulgence [Camille Flammarion]. 1867. *L'Optique*. Paris: Librairie de L. Hachette et Co.
Markley, Robert. 2005. *Dying Planet: Mars in Science and the Imagination*. Durham: Duke University Press.
Maunder, E. Walter. 1894. "The Canals of Mars." *Knowledge* 17 (11):249–252.
Maunder, E. Walter. 1895. "Section for the Observation of Mars. Report of the Section, 1892." *Memoirs of the British Astronomical Association* 2:157–198.
Maunder, E. Walter. 1903a. "A New Chart of Mars." *The Observatory* 26 (335):351–356.
Maunder, E. Walter. 1903b. "The Canals of Mars." *Knowledge* 26 (217):249–251.
Maunder, E. Walter. 1904a. "The 'Canals' of Mars. A Reply to Mr. Story." *Knowledge and Illustrated Scientific News* [New Series] 1 (4):87–89.
Maunder, E. Walter. 1904b. "The 'Canals' of Mars. A Study of an Illusion." *The Pall Mall Magazine* 34 (137):112–117.
Maunder, E. Walter. 1913. *Are the Planets Inhabited?* London: Harper and Brothers.
McCall, Sidney [Mary McNeil Fenollosa]. 1901. *Truth Dexter*. Boston: Little, Brown and Co.
McCullough, David G. 1997. *The Path Between the Seas: The Creation of the Panama Canal, 1870–1914*. New York: Simon & Schuster.
McCullough, David G. 2015. *The Wright Brothers*. New York: Simon & Schuster.
McDougall, James. 2017. *A History of Algeria*. Cambridge, U.K.: Cambridge University Press.
McGee, WJ. 1904. "Anthropology." *World's Fair Bulletin* 5 (4):4–9.
McGovern, Chauncey Montgomery. 1899. "The New Wizard of the West." *Pearson's Magazine* 7 (41):291–297.
McGrath, P. T. 1902. "Authoritative Account of Marconi's Work in Wireless Telegraphy." *The Century Magazine* 63 (5):769–782.
McIntosh, J. G. 1910. "Modern Science Notes." *Practical Teacher* 31 (1):52–53.
McKay, David S., et al. 1996. "Search for Past Life on Mars: Possible Relic Biogenic Activity in Martian Meteorite ALH84001." *Science* [New Series] 273 (5277):924–930.
McKim, Richard. 1993a. "The Life and Times of E. M. Antoniadi, 1870–1944. Part 1: An Astronomer in the Making." *Journal of the British Astronomical Association* 103 (4):164–170.
McKim, Richard. 1993b. "The Life and Times of E. M. Antoniadi, 1870–1944. Part 2: The Meudon Years." *Journal of the British Astronomical Association* 103 (5):219–227.
McKim, Richard. 2018. "Two Chessmen of Mars: Edgar Rice Burroughs and Eugène-Michel Antoniadi." *The Antiquarian Astronomer* 12:29–38.
McKim, Richard, and William Sheehan. 2009. "Schiaparelli's Final Words About Mars." *Journal of the British Astronomical Association* 119 (5):255–261.
Merriman, John M. 2009. *The Dynamite Club: How a Bombing in Fin-de-Siècle Paris Ignited the Age of Modern Terror*. Boston: Houghton Mifflin Harcourt.
Milton, Joyce. 1989. *The Yellow Kids: Foreign Correspondents in the Heyday of Yellow Journalism*. New York: Harper and Row.
Moffett, Cleveland. 1899. "Marconi's Wireless Telegraph." *McClure's Magazine* 13 (2):99–112.
Morris, Robert Griffin. 1901. *A Peep at Mars: The First "Twentieth Century Dramalet."* [Marlboro, MA]: M. Lavinia Magri.
Morrison, William. 1999. *Broadway Theatres: History and Architecture*. Mineola, NY: Dover.
Morse, Edward S. 1906. *Mars and Its Mystery*. Boston: Little, Brown and Co.
Moskowitz, Sam. 1963. *Explorers of the Infinite: Shapers of Science Fiction*. Cleveland: World Publishing.
Mousseau, Jacques. 1973. "Freud in Perspective: A Conversation with Henri F. Ellenberger." *Psychology Today* 6 (10):50–60.

Musk, Elon. 2017. "Making Humans a Multi-Planetary Species." *New Space* 5 (2):46–61.
Nall, Joshua. 2019. *News from Mars: Mass Media and the Forging of a New Astronomy, 1860–1910*. Pittsburgh: University of Pittsburgh Press.
Nasaw, David. 2000. *The Chief: The Life of William Randolph Hearst*. Boston: Houghton Mifflin.
Newcomb, Simon. 1907. "The Optical and Psychological Principles Involved in the Interpretation of the So-Called Canals of Mars." *The Astrophysical Journal* 26 (1):1–17.
Newell, Diane, and Victoria Lamont. 2011. "Savagery on Mars: Representations of the Primitive in Brackett and Burroughs." In *Visions of Mars: Essays on the Red Planet in Fiction and Science*, edited by Howard V. Hendrix, George Slusser, and Eric S. Rabkin, 73–79. Jefferson, NC: McFarland.
Newkirk, Pamela. 2015. *Spectacle: The Astonishing Life of Ota Benga*. New York: Amistad.
O'Neill, John J. 1981. *Prodigal Genius: The Life of Nikola Tesla*. Hollywood: Angriff Press.
Ohmann, Richard. 1996. *Selling Culture: Magazines, Markets, and Class at the Turn of the Century*. London: Verso.
Oppenheim, Janet. 1991. *"Shattered Nerves": Doctors, Patients, and Depression in Victorian England*. New York: Oxford University Press.
Osterbrock, Donald E. 1984. "The Rise and Fall of Edward S. Holden: Part 1." *Journal for the History of Astronomy* 15 (2):81–127.
Osterbrock, Donald E., John R. Gustafson, and W. J. Shiloh Unruh. 1988. *Eye on the Sky: Lick Observatory's First Century*. Berkeley: University of California Press.
Owen, Alex. 2004. *The Place of Enchantment: British Occultism and the Culture of the Modern*. Chicago: University of Chicago Press.
Pai, Hyung Il. 2016. "Visualizing Seoul's Landscapes: Percival Lowell and the Cultural Biography of Ethnographic Images." *Journal of Korean Studies* 21 (2):355–384.
Paine, Thomas. 1795. *The Age of Reason*. Part the First. *Being an Investigation of True and of Fabulous Theology*. London: H. D. Symonds.
Parezo, Nancy J., and Don D. Fowler. 2007. *Anthropology Goes to the Fair: The 1904 Louisiana Purchase Exposition*. Lincoln: University of Nebraska Press.
Parker, Matthew. 2007. *Panama Fever: The Epic Story of One of the Greatest Human Achievements of All Time—the Building of the Panama Canal*. New York: Doubleday.
Pascal, Blaise. 1962. *Pascal's Pensées*, translated with an introduction by Martin Turnell. New York: Harper and Brothers.
Peary, R. E. 1907. *Nearest the Pole: A Narrative of the Polar Expedition of the Peary Arctic Club in the S. S. Roosevelt, 1905–1906*. New York: Doubleday, Page and Co.
Pérez, Louis A., Jr. 1998. *The War of 1898: The United States and Cuba in History and Historiography*. Chapel Hill: University of North Carolina Press.
Pickering, William H. 1892. "Mars." *Astronomy and Astro-Physics* 11 (10):849–852.
Plank, Robert. 1968. *The Emotional Significance of Imaginary Beings: A Study of the Interaction Between Psychopathology, Literature, and Reality in the Modern World*. Springfield, IL: Charles C. Thomas.
Poole, Robert M. 2004. *Explorers House: National Geographic and the World It Made*. New York: Penguin.
Porges, Irwin. 1975. *Edgar Rice Burroughs: The Man Who Created Tarzan*. Provo: Brigham Young University Press.
Poundstone, William. 1999. *Carl Sagan: A Life in the Cosmos*. New York: Henry Holt and Co.
Powell, J. W. 1888. "From Barbarism to Civilization." *The American Anthropologist* 1 (2):97–123.
Proctor, Richard. 1881. *The Poetry of Astronomy: A Series of Familiar Essays on the Heavenly Bodies, Regarded Less in Their Strictly Scientific Aspect Than as Suggesting Thoughts*

Respecting Infinities of Time and Space, of Variety, of Vitality, and of Development. London: Smith, Elder, and Co.

Prout, Jerry. 2016. *Coxey's Crusade for Jobs: Unemployment in the Gilded Age*. DeKalb: Northern Illinois University Press.

Purbrick, Louise. 2017. "Nitrate Ruins: The Photography of Mining in the Atacama Desert, Chile." *Journal of Latin American Cultural Studies* 26 (2):253–278.

Putnam, Jennifer, and William Sheehan. 2021. "A Complicated Relationship: An Introduction to the Correspondence Between Percival Lowell and Giovanni Virginio Schiaparelli." *Journal of Astronomical History and Heritage* 24 (1):170–227.

Putnam, William Lowell, et al. 1994. *The Explorers of Mars Hill: A Centennial History of Lowell Observatory 1894–1994*. West Kennebunk, ME: Phoenix.

Raboy, Marc. 2016. *Marconi: The Man Who Networked the World*. New York: Oxford University Press.

Ramirez, Ramses M., and Robert A. Craddock. 2018. "The Geological and Climatological Case for a Warmer and Wetter Early Mars." *Nature Geoscience* 11 (4):230–237.

Rezneck, Samuel. 1953. "Unemployment, Unrest, and Relief in the United States During the Depression of 1893–97." *Journal of Political Economy* 61 (4):324–345.

Richardson, Robert S. 1954. *Exploring Mars*. New York: McGraw-Hill.

Ritchie, John, Jr. 1907. "Mars and Recent Researches." *Photo-Era* 18 (1–2):48–53, 95–102.

Robinson, Louis. 1908. "Are There Men in Other Worlds?" *Nineteenth Century and After* 63 (374):614–624.

Rogers, Alma A. 1905. "The Woman's Club Movement: Its Origin, Significance and Present Results." *The Arena* 34 (191): 347–350.

Rollyson, Carl. 2013. *Amy Lowell Anew: A Biography*. Lanham, MD: Rowman and Littlefield.

Rosebault, Charles J. 1970. *When Dana Was* The Sun*: A Story of Personal Journalism*. Westport, CT: Greenwood.

Rosenberg, Chaim M. 2011. *The Life and Times of Francis Cabot Lowell, 1775–1817*. Lanham, MD: Lexington Books.

Sagan, Carl. 1980. *Cosmos*. New York: Random House.

Sagan, Carl, and Paul Fox. 1975. "The Canals of Mars: An Assessment After Mariner 9." *Icarus* 25 (4):602–612.

Sankovitch, Nina. 2017. *The Lowells of Massachusetts: An American Family*. New York: St. Martin's.

Schiaparelli, Elsa. 1954. *Shocking Life*. New York: E. P. Dutton and Co.

Schiaparelli, Giovanni. 1894. "The Planet Mars." *Astronomy and Astro-Physics* 13 (8–9):635–640, 714–723.

Schiaparelli, Giovanni. 1898. "La Vie sur la Planète Mars." *Bulletin de la Société Astronomique de France* 12 (11):423–429.

Schiaparelli, Giovanni. 1910. "G. Schiaparelli über die Marstheorie von Svante Arrhenius." *Kosmos* 7 (8):303.

Schiaparelli, Giovanni. 1963. *Corrispondenza su Marte*. Vol. 1 (1877–1889). Pisa: Domus Galilaeana.

Schiaparelli, Giovanni. 1976. *Corrispondenza su Marte*. Vol. 2 (1890–1900). Pisa: Domus Galilaeana.

Schindler, Kevin. 2015. *The Far End of the Journey: Lowell Observatory's 24-inch Clark Telescope*. Flagstaff: Lowell Observatory.

Schwartz, A. Brad. 2015. *Broadcast Hysteria: Orson Welles's* War of the Worlds *and the Art of Fake News*. New York: Hill and Wang.

Scoles, Sarah. 2017. *Making Contact: Jill Tarter and the Search for Extraterrestrial Intelligence*. New York: Pegasus.
Scoles, Sarah. 2020. *They Are Already Here: UFO Culture and Why We See Saucers*. New York: Pegasus.
Seed, David. 2015. *Ray Bradbury*. Urbana: University of Illinois Press.
Seifer, Marc J. 1998. *Wizard: The Life and Times of Nikola Tesla: Biography of a Genius*. New York: Citadel Press.
Selley, E. A. 1907. *Astronomy for "the Man in the Street."* Dublin: Sealy, Bryers and Walker.
Serviss, Garrett P. 1889. "The Strange Markings on Mars." *The Popular Science Monthly* 35 (1):41–56.
Serviss, Garrett P. 1892a. "The Opposition of Mars." *Harper's Weekly* 36 (1861):810.
Serviss, Garrett P. 1892b. *Wonders of the Lunar World, or, A Trip to the Moon*. Urania Series, No. 1. [n.p.]: Morris Reno.
Serviss, Garrett P. 1896. "Facts and Fancies About Mars." *Harper's Weekly* 40 (2074):926.
Serviss, Garrett P. 1903. "Mars Again to the Front." *Collier's* 30 (24):17.
Serviss, Garrett P. 1904. "The Demigods of Mars." *Collier's* 33 (25):21–22.
Sheehan, William. 1988. *Planets and Perception: Telescopic Views and Interpretations, 1609–1909*. Tucson: University of Arizona Press.
Sheehan, William. 1996. *The Planet Mars: A History of Observation and Discovery*. Tucson: University of Arizona Press.
Sheehan, William. 1997. "Giovanni Schiaparelli: Visions of a Colour Blind Astronomer." *Journal of the British Astronomical Association* 107 (1):11–15.
Sheehan, William. 2015. "Mars: The History of a Master Illusionist." *The Antiquarian Astronomer* 9:14–31.
Sheehan, William. 2016. "Percival Lowell's Last Year." *The Journal of the Royal Astronomical Society of Canada* 110 (6):224–235.
Sheehan, William. 2019. "Treading Carefully: V. M. Slipher, C. O. Lampland, E. C. Slipher and Their Ambivalent Relationship with Percival Lowell's Mars." *Journal of Astronomical History and Heritage* 22 (3):365–400.
Sheehan, William. 2024. *Parallel Lives of Astronomers: Percival Lowell and Edward Emerson Barnard*. Cham, Switzerland: Springer.
Sheehan, William. 2026 [forthcoming]. "Planets, Perception, and the Canals of Mars: Do We Really Know What We See Through a Telescope?" In *Astronomy and Philosophy: Conceptual and Methodological Foundations and Challenges*, edited by Steven J. Dick. Cambridge, U.K.: Cambridge University Press.
Sheehan, William, and Jim Bell. 2021. *Discovering Mars: A History of Observation and Exploration of the Red Planet*. Tucson: University of Arizona Press.
Sheehan, William, and Thomas Dobbins. 2003. "The Spokes of Venus: An Illusion Explained." *Journal for the History of Astronomy* 34 (114):53–63.
Sheehan, William, and Stephen James O'Meara. 2001. *Mars: The Lure of the Red Planet*. Amherst, NY: Prometheus.
Sherard, R. H. 1894. "Flammarion the Astronomer." *McClure's Magazine* 2 (6):569–577.
Sheridan, Thomas E. 2012. *Arizona: A History*. Tucson: University of Arizona Press.
Shermer, Michael. 2011. *The Believing Brain: From Ghosts and Gods to Politics and Conspiracies—How We Construct Beliefs and Reinforce Them as Truths*. New York: Times Books.
Siry, Joseph M. 2002. *The Chicago Auditorium Building: Adler and Sullivan's Architecture and the City*. Chicago: University of Chicago Press.

Slipher, E. C. 1907. "Photographing Mars." *The Century Magazine* 75 (2):312.
Slipher, E. C. 1914. "The Martian Markings." *Popular Astronomy* 22 (3):155–165.
Smith, David C. 1986. *H. G. Wells: Desperately Mortal: A Biography*. New Haven: Yale University Press.
Smith, Harriette Knight. 1898. *The History of the Lowell Institute*. Boston: Lamson, Wolffe and Co.
Stephenson, Walter T. 1895. "Nikola Tesla and the Electric Light of the Future." *The Outlook* 51 (10):384–86.
Stevens, John D. 1991. *Sensationalism and the New York Press*. New York: Columbia University Press.
Stimson, Frederic Jesup. 1931. *My United States*. New York: Charles Scribner's Sons.
Stratton, Julius A., and Loretta H. Mannix. 2005. *Mind and Hand: The Birth of MIT*. Cambridge, MA: The MIT Press.
Strauss, David. 1994. "Percival Lowell, W. H. Pickering and the Founding of the Lowell Observatory." *Annals of Science* 51 (1):37–58.
Strauss, David. 2001. *Percival Lowell: The Culture and Science of a Boston Brahmin*. Cambridge, MA: Harvard University Press.
Talbot, Frederick A. 1901. "Over London in a Balloon." *The Windsor Magazine* 14 (5):495–504.
Tebbel, John. 1952. *The Life and Good Times of William Randolph Hearst*. New York: E. P. Dutton.
Terby, F. 1888. "Les Canaux de Mars, Leur Géminations et les Observations de 1888." *Ciel et Terre* 9:271–286, 289–293.
Tesla, Nikola. 1896. "Tesla's Startling Results in Radiography at Great Distances Through Considerable Thicknesses of Substance." *Electrical Review* 28 (11):131, 134–135.
Tesla, Nikola. 1899. "Some Experiments in Tesla's Laboratory with Currents of High Potential and High Frequency." *Electrical Review* 34 (13):193–197, 204.
Tesla, Nikola. 1901. "Talking with the Planets." *Collier's Weekly* 26 (19):4–5.
Tesla, Nikola. 1904. "The Transmission of Electrical Energy Without Wires." *The Electrical World and Engineer* 43 (10):429–431.
Tesla, Nikola. 1905. "The Transmission of Electrical Energy Without Wires as a Means for Furthering Peace." *The Electrical World and Engineer* 45 (1):21–24.
Tesla, Nikola. 1907. "Signalling to Mars—a Problem of Electrical Engineering." *The Harvard Illustrated Magazine* 8 (6):119–121.
Tesla, Nikola. 1919a. "My Inventions: 1. My Early Life." *Electrical Experimenter* 6 (10):696–697, 743–747.
Tesla, Nikola. 1919b. "My Inventions: 5. The Magnifying Transmitter." *Electrical Experimenter* 7 (2):112–113, 148, 173, 176–178.
Tesla, Nikola, and Aleksandar Marinčić. 1978. *Nikola Tesla: Colorado Springs Notes, 1899–1900*. Belgrade: NOLIT.
Thompson, Lawrence S. 1968. *Bibliologia Comica or Humorous Aspects of the Caparisoning and Conservation of Books*. Hamden, CT: Archon Books.
Tobin, James. 2003. *To Conquer the Air: The Wright Brothers and the Great Race for Flight*. New York: Free Press.
Todd, David. 1897. *A New Astronomy*. New York: American Book Co.
Todd, David. 1908a. "More Proof of Life on Mars." *The Circle* 3 (3):73–75, 131.
Todd, David. 1908b. "Professor Todd's Own Story of the Mars Expedition." *Cosmopolitan Magazine* 44 (4): 343–351.

Todd, David. 1920. "What If People *Do* Ridicule You!" *The American Magazine* 90 (4):48–49, 177, 180, 183–184.
Todd, Mabel Loomis. 1894. *Total Eclipses of the Sun*. Boston: Roberts Brothers.
Todd, Mabel Loomis. 1898. *Corona and Coronet: Being a Narrative of the Amherst Eclipse Expedition to Japan, in Mr. James's Schooner-Yacht Coronet, to Observe the Sun's Total Obscuration 9th August, 1896*. Boston: Houghton, Mifflin and Co.
Todd, Mabel Loomis. 1907a. "Observing an Annular Eclipse." *The Nation* 85 (2199):168–169.
Todd, Mabel Loomis. 1907b. "Photographing the 'Canals' on Mars." *The Nation* 85 (2203):264–266.
Todd, Mabel Loomis. 1908. "Our Ruddy Neighbor Planet." *The Independent* 64 (3097):791–795.
Todd, Mabel Loomis. 1912. *Tripoli the Mysterious*. Boston: Small, Maynard and Co.
Tomalin, Claire. 2021. *The Young H. G. Wells: Changing the World*. New York: Penguin.
Traxel, David. 1998. *1898: The Birth of the American Century*. New York: Alfred A. Knopf.
Tucher, Andie. 2022. *Not Exactly Lying: Fake News and Fake Journalism in American History*. New York: Columbia University Press.
Tuchman, Barbara W. 1966. *The Proud Tower: A Portrait of the World Before the War, 1890–1914*. New York: Macmillan.
Turner, H. D. T. 1980. *The Royal Hospital School, Greenwich*. London: Phillimore.
Tyndall, John. 1870. *On the Scientific Use of the Imagination*. London: Longmans, Green, and Co.
Van Biesbroeck, G. 1934. "E. E. Barnard's Visit to G. Schiaparelli." *Popular Astronomy* 42 (10):553–558.
Wallace, Alfred Russel. 1907. *Is Mars Habitable? A Critical Examination of Professor Percival Lowell's Book 'Mars and Its Canals,' with an Alternative Explanation*. London: Macmillan and Co.
Walter, Gary D. 1969. "The Korean Special Mission to the United States of America in 1883." *Journal of Korean Studies* 1 (1):89–142.
Ward, Lester F. 1907. "Mars and Its Lesson." *The Brown Alumni Monthly* 7 (8):159–165.
Ward, Lester F. 1918. *Glimpses of the Cosmos*. Vol. 6: *Period, 1897–1912. Age, 56–70*. New York: G. P. Putnam's Sons.
Waterfield, Reginald L. 1938. *A Hundred Years of Astronomy*. New York: Macmillan.
Wayman, Dorothy. 1942. *Edward Sylvester Morse*. Cambridge, MA: Harvard University Press.
Webb, George Ernest. 1983. *Tree Rings and Telescopes: The Scientific Career of A. E. Douglass*. Tucson: University of Arizona Press.
Weeks, Edward. 1966. *The Lowells and Their Institute*. Boston: Little, Brown and Co.
Weightman, Gavin. 2003. *Signor Marconi's Magic Box: The Most Remarkable Invention of the 19th Century & the Amateur Inventor Whose Genius Sparked a Revolution*. Cambridge, MA: Da Capo.
Weinersmith, Kelly, and Zach Weinersmith. 2023. *A City on Mars: Can We Settle Space, Should We Settle Space, and Have We Really Thought This Through?* New York: Penguin.
Weiss, Sara. 1905. *Journeys to the Planet Mars, or Our Mission to Ento (Mars) Being a Record of Visits Made to Ento (Mars) by Sara Weiss, Psychic, Under the Guidance of a Spirit Band, for the Purpose of Conveying to the Entoans, a Knowledge of the Continuity of Life*. Second Edition. Rochester: Austin Publishing Co.
Wells, H. G. 1898a. *Certain Personal Matters*. London: Lawrence and Bullen.
Wells, H. G. 1898b. *The War of the Worlds*. New York: Harper and Brothers.

Wells, H. G. 1908. "The Things That Live on Mars." *Cosmopolitan Magazine* 44 (4): 335–342.
Wells, H. G. 1934. *Experiment in Autobiography: Discoveries and Conclusions of a Very Ordinary Brain (Since 1866)*. New York: Macmillan.
Wells, H. G. 1937. *Star-Begotten: A Biological Fantasia*. New York: Viking.
Wells, H. G. 1998. *The Correspondence of H. G. Wells.* Vol. 1: *1880–1903*, edited by David C. Smith. London: Pickering and Chatto.
West, Anthony. 1984. *H. G. Wells: Aspects of a Life*. New York: Random House.
Westfahl, Gary. 1998. *The Mechanics of Wonder: The Creation of the Idea of Science Fiction*. Liverpool: Liverpool University Press.
Whitman, Walt. 1892. *Complete Prose Works*. Philadelphia: David McKay.
Wicks, Mark. 1911. *To Mars via the Moon: An Astronomical Story*. London: Seeley and Co.
Wilde, Oscar. 1891. *The Picture of Dorian Gray*. London: Ward Lock and Co.
Winter, Frank H. 1983. *Prelude to the Space Age: The Rocket Societies, 1924–1940*. Washington, D.C.: Smithsonian Institution Press.
Winter, Frank H. 1984. "The Strange Case of Madame Guzman and the Mars Mystique." *Griffith Observer* 48 (2):2–15.
Wright, Helen. 1987. *James Lick's Monument: The Saga of Captain Richard Floyd and the Building of the Lick Observatory*. Cambridge, U.K.: Cambridge University Press.
Yeomans, Henry Aaron. 1948. *Abbott Lawrence Lowell, 1856–1943*. Cambridge, MA: Harvard University Press.
Zubrin, Robert. 2024. *The New World on Mars: What We Can Create on the Red Planet*. New York: Diversion Books.

List of Illustrations

The illustrations in this work derive from books, magazines, newspapers, and manuscripts. Most of the captions are presented as originally written, although some have been edited for clarity and consistency of style. See "Notes on Sources" for abbreviations that refer to archival collections.

Frontispiece: Cover of *Cosmopolitan Magazine*, March 1908. Courtesy of Boston Public Library.
Page 7: Globe of Mars, from *McClure's Magazine*, March 1907. University of Colorado Boulder Libraries.
Page 10: "Harvard University, Cambridge, Massachusetts," from *Ballou's Pictorial*, January 6, 1855. Courtesy of Library of Congress.
Page 11: "Percival Lowell, Harvard class of 1876." Courtesy of Harvard University Archives, HUP, Lowell, Percival. (1A).
Page 14: "The Massachusetts Institute of Technology, Boylston Street," from *King's Handbook of Boston*, by Moses King, 1885. Courtesy of Library of Congress.
Page 14: "Edward S. Morse," from *The Sunday Herald* (Boston), February 3, 1895. Courtesy of Boston Public Library, Newspaper Collection.
Page 17: "President Arthur receiving the Corean Embassy at the Fifth Avenue Hotel," from *The Daily Graphic* (New York), September 19, 1883. Courtesy of University of Texas at Austin, Dolph Briscoe Center for American History, Newspaper Collection, camh-dob-029637.
Page 18: "The Corean Embassy to the United States," from *The Daily Graphic* (New York), September 18, 1883. Courtesy of Fenimore Art Museum Library, Cooperstown, NY.
Page 24: "Camille Flammarion," from *Les Hommes d'Aujourd'hui*, Vol. 3, no. 109, 1880. Author collection.
Page 25: "Flammarion's observatory at Juvisy," from *McClure's Magazine*, May 1894. University of Colorado Boulder Libraries.
Page 29: "Difference between the orbit of Mars and that of the Earth," from *Popular Astronomy*, by Camille Flammarion, translated by J. Ellard Gore, 1894. University of Colorado Boulder Libraries.

LIST OF ILLUSTRATIONS

Page 30: "Aspect of Mars, with its cap of polar snow," from *The Popular Science Monthly*, December 1873. Courtesy of Library of Congress.

Page 32: "The Lick Observatory, Mount Hamilton, California," from *Frank Leslie's Popular Monthly*, July 1890. University of Colorado Boulder Libraries.

Page 33: "Three strange lights in a triangle were seen through the Lick telescope," from *St. Louis Post-Dispatch Sunday Magazine*, July 7, 1907. Library of Congress.

Page 37: "Milan: Courtyard of the Palazzo di Brera," from *La Patria: Geografia dell'Italia, Provincia di Milano*, by Gustavo Strafforello and Gustavo Chiesi, 1894. Courtesy of Wikimedia Commons.

Page 38: "Giovanni Schiaparelli," from *Weltall und Menschheit*, Vol. 3, by Hans Kraemer, 1903. University of Colorado Boulder Libraries.

Page 39: "Schiaparelli's map of Mars," from *The Popular Science Monthly*, May 1889. Courtesy of Library of Congress.

Page 40: "Schiaparelli's theory caused great excitement," from *The Seattle Sunday Times*, August 22, 1909. University of Nevada Las Vegas Libraries.

Page 43: Stage set for *A Trip to the Moon*, from *The Electrical Engineer*, March 9, 1892. Courtesy of Library of Congress.

Page 44: "Garrett P. Serviss," from *The Columbian* (Bloomsburg, PA), January 13, 1898. Courtesy of Andruss Library, Commonwealth University–Bloomsburg.

Page 49: "Flagstaff, Arizona, and the San Francisco Peaks," inset from lithograph *The Grand Canyon of the Colorado River, Arizona*, by Jules Baumann, 1892. Courtesy of www.oldmapsofthe1800s.com.

Page 50: "The observatory of Mr. Percival Lowell of Boston at Flagstaff, Arizona," from *The Sunday Post* (Boston), August 30, 1896. Courtesy of Lowell Observatory Archives, PLP, Lowell Scrapbook 1894–1897.

Page 53: "Mars. Nov. 4, 1894. (P. Lowell)." Courtesy of Lowell Observatory Archives, HLLO, Logbook 2.

Page 54: "Mars. Nov. 19, 1894. (P. Lowell)." Courtesy of Lowell Observatory Archives, HLLO, Logbook 3.

Page 59: Lowell Institute advertisement for lectures on "The Planet Mars," from *Boston Daily Advertiser*, January 17, 1895. Courtesy of Boston Public Library, Newspaper Collection.

Page 60: "Mr. Percival Lowell," from *The Sunday Herald* (Boston), January 7, 1900. Courtesy of University of Arizona Libraries, Special Collections, AEDP, Box 16, Folder 2.

Page 62: "The planet Mars," from the Lowell Observatory, 1894. University of Arizona Libraries, Special Collections, AEDP, Box 16, Folder 6.

Page 65: "Lilian Whiting," from *The Burlington Hawk-Eye* (Burlington, IA), October 29, 1899. Courtesy of Burlington Public Library.

Page 69: "The drawing-room in M. Flammarion's Paris home," from *McClure's Magazine*, May 1894. University of Colorado Boulder Libraries.

Page 71: "Map of Mars on Mercator's projection," from *Mars*, by Percival Lowell, 1895. Author collection.

Page 72: "Cassini's sketch of Mars," from *Martis Circa Axem Proprium Revolubilis Observationes Bononiae*, by Giovanni Domenico Cassini, 1666. Courtesy of University of Oklahoma Libraries, History of Science Collections.

LIST OF ILLUSTRATIONS 307

Page 74: "A summer scene on Mars," from *The Sunday Post* (Boston), May 31, 1896. Courtesy of Boston Public Library, Newspaper Collection.

Page 76: "Chart of visible hemisphere of Venus (P. Lowell)," from *A New Astronomy*, by David P. Todd, 1897. University of Colorado Boulder Libraries.

Page 78: "Hamilton Hotel," from *Stark's Illustrated Bermuda Guide*, by James H. Stark, 1897. University of Texas at Austin Libraries.

Page 80: "H. G. Wells," from *The War of the Worlds*, by H. G. Wells, 1898. Colorado State University Libraries.

Page 81: Bicyclist on a country road, from *The Wheels of Chance*, by H. G. Wells, 1896. Courtesy of Library of Congress.

Page 83: "The War of the Worlds by H. G. Wells," from *The Cosmopolitan*, April 1897. Courtesy of Library of Congress.

Page 83: Martian tripod hunting humans, from *Pearson's Magazine*, June 1897. Courtesy of Library of Congress.

Page 84: "The actual Martians were the most extraordinary creatures it is possible to conceive," from *Pearson's Magazine*, November 1897. Courtesy of Library of Congress.

Page 86: "The destruction of Brooklyn Bridge," from *The New York Evening Journal*, December 24, 1897. Library of Congress.

Page 90: "Nikola Tesla," from *The Electrical World*, August 4, 1894. Courtesy of Library of Congress.

Page 93: "The Auditorium Building and Lake Front Park," from *Auditorium*, by Edward R. Garczynski, 1891. Courtesy of Library of Congress.

Page 96: "Experimental station at Colorado Springs," from *The New York Herald*, November 12, 1899. Library of Congress.

Page 98: "David and Mabel Todd, Amherst, Massachusetts." Courtesy of Yale University Library, TBPC, Box 4, Folder 56.

Page 103: Newsboys hawking news of the Martians, from *The Sunday Post* (Boston), January 16, 1898. Courtesy of Boston Public Library, Newspaper Collection.

Page 106: Front page of *The Examiner* (San Francisco), January 4, 1901. Courtesy of San Francisco Public Library, Magazines and Newspapers Center.

Page 109: Cover of *The New Golden Hours*, March 30, 1901. Courtesy of Tampa Library, University of South Florida, Special Collections.

Page 110: "The first message from Mars," from *Life*, July 4, 1901. Courtesy of Library of Congress.

Page 111: "A Signal from Mars," sheet music. Courtesy of New York Public Library.

Page 111: "Mrs. Tom Thumb: The little lady." Courtesy of Barnum Museum, Bridgeport, CT, Accession no. EL 1988.106.001.

Page 112: "He meets the man from Mars," from *The Morning Telegraph* (New York), April 5, 1903. University of Nevada Las Vegas Libraries.

Page 116: "Sir Robert S. Ball," from *Vanity Fair*, April 13, 1905. Courtesy of Yale Center for British Art.

Page 118: "Signal Hill, St. John's, from the sea," from *Harper's Weekly*, January 4, 1902. Courtesy of University of Colorado Boulder Libraries.

Page 120: "All is vanity," from *Life*, November 27, 1902. University of Colorado Boulder Libraries.

Page 122: "Marching to the dining hall," from *The Illustrated London News*, February 19, 1848. Courtesy of Denver Public Library.

LIST OF ILLUSTRATIONS

Page 124: "Now it's up to Tesla to connect us with Mars," from *The Brooklyn Daily Eagle*, December 23, 1902. Courtesy of Brooklyn Public Library, Local Newspapers on Microfilm Collection, Center for Brooklyn History.

Page 125: Tesla's tower at Wardenclyffe, from *Boston American*, June 13, 1909. Courtesy of Boston Public Library, Newspaper Collection.

Page 125: "Key map," from *Monthly Notices of the Royal Astronomical Society*, June 12, 1903. Courtesy of University of Arizona Libraries, Special Collections, AEDP, Box 16, Folder 5.

Page 125: "Drawing submitted to the boys," from *Monthly Notices of the Royal Astronomical Society*, June 12, 1903. Courtesy of University of Arizona Libraries, Special Collections, AEDP, Box 16, Folder 5.

Page 126: "Drawings made by boys placed at the indicated distances from the original drawing," from *Monthly Notices of the Royal Astronomical Society*, June 12, 1903. Courtesy of University of Arizona Libraries, Special Collections, AEDP, Box 16, Folder 5.

Page 130: "View from the upper floors of the Flatiron Building," from *Munsey's Magazine*, July 1905. Courtesy of Denver Public Library.

Page 132: "Experimental séance at the apartment of Camille Flammarion. Rue Cassini, Paris, 1898." Courtesy of Wikimedia Commons.

Page 135: "Detail from Lowell Observatory Mars map," from Percival Lowell's hand-drawn map of Mars, 1905. Courtesy of Lowell Observatory Archives.

Page 136: "North tropic canals," from *Lowell Observatory Bulletin No. 12*, 1904. University of Colorado Boulder Libraries.

Page 137: "Mean canal cartouches," from *Lowell Observatory Bulletin No. 12*, 1904. University of Colorado Boulder Libraries.

Page 142: "Camille Flammarion and Percival Lowell at Juvisy," from *L'Astronomie*, May 1930. Courtesy of Lowell Observatory Archives.

Page 144: "World's Fair, St. Louis, 1904." Courtesy of Missouri Historical Society, N16107.

Page 149: "Edward S. Morse at Lowell's telescope." Courtesy of Lowell Observatory Archives, HPLO, 0116a.

Page 150: "Photograph of Mars by C. O. Lampland," from *Lowell Observatory Bulletin No. 21*, 1905. University of Colorado Boulder Libraries.

Page 152: "African village, St. Louis World's Fair." Courtesy of Missouri Historical Society, N16492.

Page 154: "How various modern scientists have figured out that a dweller on the planet Mars would probably look," from *The San Francisco Examiner*, March 18, 1906. Courtesy of California State Library, California History Room.

Page 156: "Mars people as they may look," from *The Telegraph-Herald* (Dubuque), November 2, 1907. Courtesy of State Historical Society of Iowa.

Page 159: "Professor Percival Lowell," from *The Saturday Evening Mail*, February 9, 1907. Author collection.

Page 161: "There Is Life on the Planet Mars," *The New York Times*, December 9, 1906. University of Colorado Boulder Libraries.

Page 164: "Robert and Josephine Peary," from *Peary*, by Fitzhugh Green, 1926. University of Colorado Boulder Libraries.

LIST OF ILLUSTRATIONS   309

Page 166:   "Lowell Expedition to the Andes MCMVII," letterhead, David Todd to Percival Lowell, May 1, 1907. Courtesy of Lowell Observatory Archives, PLP, Percival Lowell Correspondence.
Page 167:   "The Latest from Mars," from *The New York Herald*, November 28, 1909. Library of Congress.
Page 169:   "David and Mabel Todd with the Amherst telescope at Alianza, Chile." Courtesy of Yale University Library, TBPC, Box 42, Album 17.
Page 172:   "Dancing the cueca," from *The Graphic* (London), April 17, 1880. Courtesy of San Francisco Public Library, Magazines and Newspapers Center.
Page 175:   "Waldorf-Astoria Hotel, New York," from *Architecture and Building*, February 5, 1898. Courtesy of New York State Library.
Page 176:   "Taken July 13, when Mars was nearest the Earth," from *The Century Magazine*, December 1907. University of Colorado Boulder Libraries.
Page 181:   "Martians gazing at the Earth," from *The Sphere*, July 13, 1907. Courtesy of Library of Congress.
Page 187:   "Mr. Skygack, from Mars," from *The San Diegan-Sun*, November 13, 1907. Courtesy of California State Library, California History Room,.
Page 188:   "'There are certain features in which they are likely to resemble us . . .'" from *Cosmopolitan Magazine*, March 1908. Courtesy of Library of Congress.
Page 189:   "Lecture! 'Latest News from Mars,'" from *The Racine Daily Journal*, March 16, 1908. University of Wisconsin Library.
Page 193:   "Wilbur Wright's aeroplane in flight near Le Mans," *Le Petit Journal*, August 30, 1908. Author collection.
Page 194:   "Mr. and Mrs. Percival Lowell." Courtesy of Lowell Observatory Archives, HPLO, 0115.
Page 196:   "Hyde Park and the Serpentine," from *Mars as the Abode of Life*, by Percival Lowell, 1908. University of Colorado Boulder Libraries.
Page 197:   "Aerial Houseboats and Villages of Mars," from *The New York Herald*, August 15, 1909. New York State Library.
Page 199:   "Professor Pickering's plan for communicating with Mars," from *The New York Herald*, April 25, 1909. Library of Congress.
Page 200:   "Mars to be signaled by scientist and aeronaut," from *The San Francisco Examiner*, May 25, 1909. Courtesy of San Francisco Public Library, Magazines and Newspapers Center.
Page 202:   "'Just off!'" Courtesy of Yale University Library, TBPC, Box 144, Folder 930.
Page 205:   "Mr. Skygack, from Mars," from *San Antonio Light and Gazette*, September 21, 1909. Courtesy of University of Texas at Austin, Dolph Briscoe Center for American History, Newspaper Collection, camh-dob-029365.
Page 207:   "Yerkes Observatory," from *Beautiful Lake Geneva*, by N. W. Smails, 1895. Courtesy of Lake Geneva Public Library.
Page 210:   "Le Petit Journal: La conquête du pole nord," from *Le Petit Journal*, September 19, 1909. Author collection.
Page 211:   "Mr. E. M. Antoniadi," from *The British Chess Magazine*, September 1907. Courtesy of Library of Congress.
Page 212:   "Observatoire de Meudon," from *Annales de l'Observatoire d'Astronomie*

Physique de Paris Sis Parc de Meudon (Seine-et-Oise), Vol. 1, Plate 9, 1896. Courtesy of Library of Congress.

Page 215: "To Mr. Percival Lowell, Splendid definition," enclosure with E. M. Antoniadi to Percival Lowell, October 9, 1909. Courtesy of Lowell Observatory Archives, PLP, Percival Lowell Correspondence.

Page 215: "To Mr. Percival Lowell, Tremulous definition," enclosure with E. M. Antoniadi to Percival Lowell, October 9, 1909. Courtesy of Lowell Observatory Archives, PLP, Percival Lowell Correspondence.

Page 216: "Mars. Oct. 31; 1909. (P. Lowell)." Courtesy of Lowell Observatory Archives, HLLO, Logbook 21.

Page 219: "Sion College, new building on the Thames Embankment," from *The Illustrated London News*, October 9, 1886. Courtesy of Denver Public Library.

Page 220: "Views of Mars (Syrtis Major and Lacus Moeris region) in 1909 with various telescopes," from *The Journal of the British Astronomical Association*, April 1910. (Caption is adapted from reprint of image in *Knowledge*, May 1913.) Courtesy of Lowell Observatory Archives.

Index

Page numbers in italic refer to illustrations and accompanying captions. Page numbers after 254 refer to endnotes.

Adams, John, 10
Adams, Samuel, 10
Addams, Jane, 157
advertising, 110, *110*
Aero Club of Great Britain, 194
Aero Club of New England, 194, 199, 202
Age of Reason, The (Paine), 185
air, 27, 44, 48, 53, 60, 61, 66, 92, 100, 146, 155, 193, 209, 239; *see also* atmosphere(s)
aircraft, experimental, 109
Alaska, 174, 204–5
Albuquerque, New Mexico, 157
alchemy, 183–84
Algeria, 73–74
Alianza, Chile, 168–72, *169*, 174, 178–79, 220; *see also* Lowell Expedition to the Andes
"All Is Vanity" (visual illusion), 120, *120*, 127
Alta Vista Hotel (Colorado Springs), 95
alternating current, 92
American Astronomical Society, 206
American Museum of Natural History, 177
American Red Cross, 100, 101, 106
American Revolution, 184–85
Amherst College, 97, 99, 165, 169, 171, 179, 199, 232
Amherst, Massachusetts, 97–98, *98*, 99
anarchists, 22, 31, 54–55, 67, 97
Andes Mountains, 165, 168; *see also* Lowell Expedition to the Andes
Antarctica, 239
Anthony, Susan B., 101
anthropology, 21, 153, 224

Anthropology Villages (St. Louis World's Fair), *152*, 152–53
anthropomorphism, 79, 162
anti-canalists, 72, 118, 128, 170, 184, 263
Antoniadi, Eugène-Michel, 76, 210–17, *211*, *215*, 219, 220, *220*, 263, 270
Apache tribe, 48, 145
Apollo program, 3
Arapaho tribe, 95
Arctic exploration, 163–64, 173, 177, 204–5, *205*, 209–10, *210*, 217–18, 221
Argentina, 227
Arizona, 47–52, 70, 75, 111, 113–15, 138, 145, 147, 148, 151, 155, 157–58, 177, 214, 229; *see also specific locations, e.g.:* Flagstaff
Armory Show (New York, 1913), 224
Armstrong, Neil, 3
Arnold, Edwin, 55–56
arsenic (as a treatment for neurasthenia), 77
Arthur, Chester A., *17*, 17–18
Arthur, King (legendary character), 58, 67, 68, 139
assassinations, 22, 54–55, 97, 226
Associated Press, 41, 204
Astronomical and Astrophysical Society of America, 206–8
Astronomical Society of France, 23, 276
astronomy, 12, 23, 27, 36, 41, 44, 51, 55–56, 66, 69, 116, 129, 191, 208, 235, 236, 282
Atacama Desert, 168, 178
atheists, 80, 185
Atlantic Monthly, The, 10, 20, 65–67, 113

atmosphere(s)
  Earth's, 49, 53, 70, 74, 95, 96, 107, 143, 144, 166, 170, 178, 194, 199, 201, 212, 215–16
  Martian, 60, 66, 155, 209, 239, 240
  planetary, 27
Auditorium Building (Chicago), 92, *93*, 113–14
automobiles, 97, 109, 164, 192, 202
aviation, 192–93, *193*, 196–97, 199

Babylonians, 29
Back Bay (Boston neighborhood), 14, 15, 116
Bacon, Francis, 230
Ball, Sir Robert, *116*, 116–17, 151, 154, *154*
ballet, 223–24
balloon flights, 24, 194–96, *202*, 202–3, 280
"barbarians," 21, 224
Barcelona, Spain, 54
Barnum, P. T., 111
"Baronial Mansion," 122, 151, 164, 170, 214
Belgium, 226
Bell, Alexander Graham, 2, 177, 184
Bermuda, 77, *78*
Bernhardt, Sarah, 35, 43
Berson, Arthur, 200, 201
Bible, 12, 46, 185
bicycles, 81, *81*
Binghamton, New York, 2
Bloomsbury Group, 223
bombings, 22, 31, 54
Boston (airship), *202*, 202–3
Boston, Massachusetts, 9, 13–16, *14*, 19, 20, 21, 35, 43, 45, 47–48, 51, 58, 65, 68, 75, 77, 81, 87, 92, 97–100, 113, 116, 117, 122, 131, 139, 144, 151, 157, 159, 162–64, 170, 177, 186, 189, 194, 221, 224, 225
Boston American, 170
Boston Brahmins, 10, 12–13, 16, 18, 19, 77, 92
Boston Commonwealth, 64
Boston Daily Globe, The, 41, 99, 160
Boston Evening Transcript, 15, 35, 158–59, 208
Boston Herald, The, 48, 60
Boston Post, The, 87, 157, 235
Boston Public Library, 58, 59, 68, 87
Boston Society of Natural History, 35
Boston Sunday Globe, The, 68
Boston *Sunday Herald*, 62, 156
Boston *Sunday Post*, 74, 74–75, 198
Brackett, Leigh, 236
Bradbury, Ray, 236, 238
Brera Observatory (Milan), 37–38, 54
Brisbane, Arthur, 86, 87, 91, 107, 155
Britain, 26, 58, 82, 92, 184, 226, 227; *see also* England
British Astronomical Association, 127, 151, 217–21
Bronx Zoo, 158
Brooklyn, New York, 44, 130, 189
Brooklyn Bridge, 86, *86*
Brooklyn Citizen, The, 68
Brooklyn Daily Eagle, The, 44
Brownie cameras, 143, 153
Brown University, 184
Bryant Park (New York), 232
Bugs Bunny, 3
Burnham, S. W., 88–89
Burroughs, Edgar Rice, 236, 247

California, 3, 227, 237
California Academy of Sciences, 28
Calvary Episcopal Church (New York), 130–1, 140
Cambridge, Massachusetts, *10*, 149, 189
Cambridge University, 116
cameras, 143
Campbell, William Wallace, 203, 285
canalists, 72, 117–18, 128, 151, 155–56, 161, 210, 238, 263
canals, Martian
  Eugène-Michel Antoniadi on, 210–11, 213–17, *215*, 219–20, *220*
  appeal of belief in, 227, 230
  Robert Ball on, 117, 151
  double canals, 39, 45, 54, 73, 75, 148, 165, 171–72, 173–74, 189
  Camille Flammarion on, 70–72, 150–51, 223
  Percival Lowell and, 47, 51, 53–54, 62–64, 70–71, *71*, 75, 115, 123, 134–38, *135*, 144, 147–49, 154, 170–72, 176–77, 209, 215–17, *216*, *220*, 225, 227–28

INDEX

and Lowell's legacy, 234–38
Edward Walter Maunder on, 121,
  125–28, 143, 171, 219–21
photographs of, 2, 144, 149–51, *150*,
  156, 160, 171–74, *176*, 176–77, 203,
  207, 211
in the press, 2, *40*, 40–41, 44–45, 81,
  88–89, 113, 117, 125, 149, 156, 158–59,
  167, *167*, 172, 173–74, 194, 221, 225
Giovanni Schiaparelli on, *39*, 39–40, 44,
  54, 73, 141, 221, 222–23
Garrett P. Serviss on, 44–45, 87, 128,
  156, 222, 234
skepticism about, 41–42, 47, 66, 72,
  88–89, 115, 118, 121, 125–29, 143,
  170–71, 184, 203, 209, 213–17,
  219–20, 222–23, 282
and "Small Boy Theory," *125*, 125–29,
  126, 143, 151, 211, 219
Nikola Tesla on, 174, 223
David Todd on, 97, 172–74, 191, 208,
  222, 282
Mabel Todd on, 170, 173, 189
Canal Zone, 167
capitalism, 22
Carnegie, Andrew, 31
Carnot, Sadi, 54–55
cartouches, *136*, 136–37, *137,* 141, 160
Cassini, Giovanni Domenico, 69, 71–72,
  72, 76
Central Africa, 62, 68, 153
Century Magazine, The, 174, *176,* 176–77,
  184, 186
Cheyenne tribe, 95
Chicago, Illinois, 2, 31, 66, 88, 92–95, *93,*
  97, 100, 113–14, 157, 189, 206
Chicago Daily Tribune, The, 23, 194
Chicago Sunday Tribune, The, 108
Chicago World's Fair (1893), 31
Chile, 168, 171, 172, 174, 177, 178, 189,
  190, 203; *see also specific locations, e.g.:*
  Alianza
Chinese space agency, 240
Christianity, 12, 185
Christian Work and the Evangelist, The, 185
Christmas Carol, A (Dickens), 113
clairvoyance, 131
Clarke, Arthur C., 236, 237–38
class warfare, 22, 55

Clerke, Agnes, 67, 271
Cleveland Plain Dealer, 231
Clinton, Bill, 239
clock drives, 37–38
Colorado, 95, 106, 147
Colorado Springs, Colorado, 95–96, *96,* 266
Columbia University, 86, 174
Columbus, Christopher, 31, 193
comets, 27, 38, 143
Common Sense (Paine), 184–85
Como, Lake, 141
Conan Doyle, Arthur, 233
"Concerning the Nose" (Wells), 80
Condo, A. D., 186
confirmation bias, 230
Conrad, Joseph, 62
Cook, Frederick A., 204–5, 210, *210,* 218, 220
Copenhagen, Denmark, 218
Copley Square (Boston), 58, 159
Cornell University, 114, 145
Cornwall, England, 123
Cosmopolitan, The (later called *Cosmopolitan Magazine*), 82, 187, 236
Coxey, Jacob, 55
Crane, Stephen, 86
Croll, James, 35
Cuba, 85, 87, 88
Cubists, 224
cueca (dance), 172, *172*
Culebra (Panama), *167,* 168, 276–77
Curie, Marie, 224
Curie, Pierre, 224
Curtis, Ralph Wormeley, 284
"cutting," 16

Darwin, Charles, 12, 57, 80, 83, 155,
  195–96, 209
Davidson, George, 28
Dayton, Ohio, 193
Debs, Eugene V., 55
Denver, Colorado, 95
depression (psychological disorder), 77, 225
depression of 1893, 55
Diablo Range (California), 32
Dickinson, Austin, 98
Dickinson, Emily, 97–98
direct current, 92
Dom Pedro II, Emperor of Brazil, 25

double canals, 39, 45, 54, 73, 75, 148, 165, 171–72, 173–74, 189
Douglass, Andrew Ellicott, 48–50, 51, 107, 114–16, 134, 139, 170–71
Duluth, Minnesota, 83, 111
"dumbbell," Mars, 72, *72*, 76

Earp, Wyatt, 49
Earth
    atmosphere of, 49, 53, 70, 74, 95, 96, 107, 143, 144, 166, 170, 178, 194, 199, 201, 212, 215–16
    orbit of, compared with Mars, 29, *29*–30
    origin of life on, 239–40
    as seen from space, 43, *43*
Eastman Kodak Company, 143, 153
eclipses
    lunar, 38
    solar, 98–99, 165
Ecuador, 168
Edison, Thomas, 87, 88, 92, 120, 192
Edison's Conquest of Mars (Serviss), 87, 91, 235
Edwardian England, 195, 223
Edward VII, King, 194
Egypt, 63, 135
Eiffel Tower, 119–20, 212, 225
Einstein, Albert, 225
Eldridge, Helen, 66
electricity, 91, 92, 96
Ends of the Earth Club, 164
England, 12, 13, 58, 72, 107, 194, 217; *see also* Britain
English Channel, 92
Enlightenment, 26
evolution, theory of, 12, 20–21, 57, 80, 83, 84, 154–56, 209
extraterrestrial life, notions of, 26, 185; *see also* Martians

Fenollosa, Mary McNeil, 269
Ferguson, Charles, 186
Fifth Avenue Hotel (New York City), 17, *17*
fin-de-siècle, 22, 55
Fish, Mamie, 1
Flagstaff, Arizona, *49*, 49–50, *50*, 52, 54, 56–57, 75, 77, 107, 114–15, 122–23, 133, 139, 141, 147–51, 157, 165, 170–72, 214, 216, 227–29; *see also* Lowell Observatory

Flammarion, Camille, 21, 23–28, *24*, *25*, 31, 33, 36, 40–42, 45, 46, 57, 60, 66, *69*, 69–73, 76, 81, 94, 100, 101, 119, 128–29, 131, 132, *132*, 141, 142, *142*, 150–51, 165–66, 190–91, 201, 210, 223, 225, 233, 256–57, 276
Flammarion, Sylvie, 23–24, 70, 256
Flatiron Building (New York), 130, *130*, 192
Flournoy, Théodore, 132–33
Ford Motor Company, 192
Fort Myer, Virginia, 197
France, 22–23, 26, 40, 54–55, 69–70, 73–74, 97, 142, 193, *193*, 197, 210, 213, 217, 226
Franklin, Benjamin, 26
Franz Ferdinand, Archduke, 226
free verse *(vers libre)*, 224, 225
French Academy of Sciences, 217
Freud, Sigmund, 133
Futurists, 224

Garrick Theatre (New York), 112
gemination. *See* double canals
Geneva, Lake (Wisconsin), 206–8
Gentola (alleged Martian woman), 131
Germany, 178, 226, 227
Gernsback, Hugo, 235
Geronimo, 48, 145
Gilded Age, 12, 15, 20
God, 55, 185, 191
Goddard, Robert H., 235, 287
Gold Rush (California), 32
Gramercy Park (New York), 130
Grand Canyon, 49, 148, 151
Grand Hôtel Costebelle (Hyères, France), 97
gravity, 138, 154
Greece, 217
Greenland, 205, 218
Greenwich, England, 120–21, 128; *see also* Royal Observatory (Greenwich, England)
Gregory, Richard, *154*, 155
Guam, 88
Guzman, Anne-Émile-Clara Goguet, 22–23, 28, 106–7, 191, 206

Haggard, H. Rider, 62
Hale, Edward Everett, 64
Hamilton, Edmond, 236

INDEX

Hamilton Hotel (Bermuda), 77, *78*
Hancock, John, 10
Harper's New Monthly Magazine, 68
Harper's Weekly, 44
Hartford Courant, The, 231
Harvard Alumni Association, 10
Harvard College (Harvard University), 9–13, *10, 11,* 16, 18, 31, 40, 45, 47–48, 51, 61, 85, 113, 114, 184, 194, 198, 224, 225, 229
Harvard College Observatory, 31, 40, 45, 47, 148, 149, 207
Havana, Cuba, 87
Hawthorne, Julian, 108, 266
Hawthorne, Nathaniel, 108
Hawtrey, Charles, 268
Haymarket riot, 31
Hearst, William Randolph, 85–88, 91, 94, 100, 105, 106, 107, 117, 142, 153, 155, 170, 173–74, 193
Henry, Lord (fictional character), 22
Herschel, William, 26
Hodgson, J. Hamilton P., 1
Holden, Edward S., 41, 51–52, 66, 72, 76, 78, 108
Holmes, Oliver Wendell, Sr., 12
Holy Grail, 58, 139
Homestead Steel Works, 31
Hong Yong Sik, *18*
Hugo Awards, 235
Hull House (Chicago), 157
Hussey, William J., 264, 282
Huxley, Thomas Henry, 80
Hyde Park (London), 196, *196*
Hyères, France, 97

ice caps, Martian. *See* polar ice caps (Mars)
Illinois, 64, 218
imperialism, 58, 88, 145
India, 131
industrialization, 22
Industrial Revolution, 13
Iquique, Chile, 168, 178
Ireland, 185
Italy, 92, 141–42

Japan, 14–15, 17, 18, 20–21, 35, 42, 45, 49, 56, 59, 117, 134, 138, 147
Jet Propulsion Laboratory, 237
Jim Crow laws, 109–10
Johns Hopkins University, 198
Jones, Jane, 2
Juneau, Alaska, 174
Jupiter, 26, 27, 193
Juvisy, France, 24–26, *25*, 76, 142, *142*, 150, 210

Kansas City, Missouri, 186
Keller, Helen, 145
Kepler, Johannes, 26
King Solomon's Mines (Haggard), 62
Kingston, Jamaica, 191
Kipling, Rudyard, 62
Knickerbocker Trust Company, 173
Knights of the Round Table, 58, 139
Korea and Koreans, *17,* 17–20, *18,* 28, 138
Kuiper, Gerard, 234

Ladies' Home Journal, The, 109
Lake Front Park (Chicago), *93*
Lampland, Carl Otto, 144, *150*, 160
Lang, Andrew, 67
Latham, Philip (pseudonym of Robert S. Richardson), 236
Lawrence, Massachusetts, 13, 19
Le Bovier de Fontenelle, Bernard, 26
Lee, Rose, 16, 138
Le Mans, France, 193, *193*
Leuschner, Armin Otto, 285
Library of Congress, 109
Lick, James, 32
Lick Observatory, 32, *32,* 33, 40, 41, 51, 66, 72, 76, 78, 108, 203, 209, 216, 228
life, origin of, 239–40
Life magazine, 120
Lilian Whiting Club, 64
literature, 62, 223; *see also* science fiction
"Living Things That May Be, The" (Wells), 80
London, England, 21, 54, 79, 83–84, 86, 112, 121, 125, 128–29, 194–96, *196,* 218–20, *219,* 221, 226
Long Island (New York), 2, 124, 146, 147, 175, 232
Los Angeles, California, 170, 172, 203
Los Angeles Herald (also *Daily Herald*), 107, 221
Los Angeles Times, The, 162
Louisiana Purchase Exposition. *See* St. Louis World's Fair (1904)

Louis XIV, King, 69, 211
Lovecraft, H. P., 236
Lowell, A. Lawrence, 116, 122, 224, 225
Lowell, Amy, 224–26, 228
Lowell, Augustus, 10, 13, 15–16, 59, *59*, 75, 77, 116, 122, 138
Lowell, Constance Savage Keith, 194–95, *194*, 228, 229, 231
Lowell, Elizabeth. *See* Putnam, Elizabeth Lowell
Lowell, James Russell, 10
Lowell, Katharine. *See* Roosevelt, Katharine Lowell
Lowell, Massachusetts, 13, 19
Lowell, Percival, *11, 60, 159*
  amorous affairs of, 138–40
  and Eugène-Michel Antoniadi, 211–17
  and Robert Ball, 116–17
  as canalist, 72, 117–18, 127
  and cartouches of Mars, *136*, 136–37, *137*, 141, 160
  death of, 228–29
  declining public support for and attacks on, 206–9, 221, 222–23
  and A. E. Douglass, 48–51, 114–16, 139, 170–71
  in England, 68, 72–73, 194–96, *196*
  in Europe, 73, 141–42, 225–26
  and Expedition to the Andes, 165, 166, 168, 170–72, 174
  family of, 10, 12–16, 35, 224, 226
  in Far East, 17, 19–22, 35, 42, 45, 56, 134, 138
  in Flagstaff, 46–47, 52–54, 56, 57, 75, 114–15, 122, 123, 138–39, 147–49, 151, 170, 172, 214–16, 227, 228
  and Camille Flammarion, 45, 46, 70–72, 76, 128, 141, 142, *142*, 150, 165–66, 223, 225
  in France, 70–72, 142, *142*, 194, 225
  growing interest in astronomy, 45–47
  and Korean delegation, *17*, 17–19, *18*
  as lecturer and public speaker, *59*, 59–64, 68, 133–34, 159–60, 177, 227–28, 275
  legacy of, 229–31, 233–41
  marriage of, *194*, 194–95
  Mars, 68, 73, 117, 235, 287
  Mars and Its Canals, 158, 214, 236
  Mars as the Abode of Life, 186, 196, 287
  on Martians, 154, *154*, 158, 160, *161*, 228
  mental collapse of, 75–79, 91, 97, 99–100, 113, 225
  and Edward Morse, 15, 16–17, 35, 147–48, *149*, 151, 155–56, 227, 276
  in North Africa, 73–74
  on Robert E. Peary, 205
  personality of, 133, 134, 138, 284–85
  philanthropy of, 122
  and photographs of Mars, 143–44, 149–51, 156, 160, 174, *176*, 176–77, 184, 186, 203, 207, 211, 214, 219–20
  and William H. Pickering, 47–48, 51, 198, 259
  as poet, 56, 57
  public opinion in favor of, 184–86, 191, 203
  on role of imagination in science, 42
  and Theodore Roosevelt, 16, 34, 85
  and Giovanni Schiaparelli, 73, 77–78, 141, 158, 165, 222–23
  and Garrett P. Serviss, 67, 128, 142, 156, 229
  sketches of Mars, *53*, 54, *54*, *62*, 134–36, *135*, 143, *176*, 177, 211, *216*, *220*, 225
  sketch of Venus, *76*, 76–77, 113, 114, 127
  and Nikola Tesla, 114, 174, 223
  and David and Mabel Todd, 97–100, 165, 166, 172, 189, 203, 208–9
  and H. G. Wells, 80, 156–57, 187–88
  and Lilian Whiting, 64–66, 114, 160–61, *161*, 233, 275, 287
Lowell Expedition to the Andes, 165–74, *166, 169,* 177, 189–90; *see also* Alianza, Chile
Lowell Institute, 14, 59, *59*, 65, 68, 116, 159, 261, 287
Lowell Observatory (Flagstaff, Arizona), 50, *50*, 51, 52, 53, 56–57, 70, 75, 107, 114, 122, 135, *135*, 147, *149*, 149–50, 156, 164, 170–71, 198, 207, 209, 219–20, *220*, 225, 227; *see also* Mars Hill
Lucerne, Switzerland, 142
Lumière brothers, 120
lunar eclipses, 38
Luxembourg, 235
Lyon, France, 54

INDEX

Mackay, Charles, 183–84
Madison Square Garden, 265–66
Madrid, Spain, 31
Magri, Lavinia (Mrs. Tom Thumb), 111, *111*
Magri, Primo, 111
Mamaroneck, New York, 34
Marconi, Guglielmo, 92, 107–8, 118, 123–24, *124*, 146, 176
Mariner 9 spacecraft, 237, 238
Marion, Fulgence (pseudonym of Camille Flammarion), 119, 129
Mars
    Antoniadi's sketches of, at Meudon, 214–17, *215*, *220*
    at astronomical opposition, *29*, 29–31, 39, 47, 107, 123, 142, 144, 165, 191, 198, 204, 211, 233
    atmosphere of, 60, 155, 209, 239, 240
    canals on. *See* canals, Martian
    Cassini's "dumbbell" sketch of, 72, *72*, 76
    Lowell's sketches of, 53, 54, *54*, *62*, 134–36, *135*, 143, *176*, 177, 211, *216*, *220*, 225
    maps of, 38–39, *39*, 45, 70–71, *71*, 125, *125*, *135*, 211, 238
    orbit of, compared with Earth, *29*, 29–30
    photographs of, 144, 149–51, *150*, 160, 171–74, *176*, 176–77, 186, 203, 207, 211, 214, 219–20, *220*, 222, 237
    polar ice caps of, 30, *30*, 31, 47, 52–53, 57, 123, 136, 137, 150, 158, 162, 164, 170, 241
    proposed communication with, 2, 27–28, 94–95, 96, *124*, 175, 195, 198–99, *199*, 199–200, *200*, 233
Mars (P. Lowell), 68, 73, 117, 235, 287
Mars and Its Canals (P. Lowell), 158, 214, 236
Mars as the Abode of Life (P. Lowell), 186, 196, 287
Mars Hill, 47, 50, 52, 56, 114, 122, 148, 170, 214, 227, 228–29; *see also* Lowell Observatory
Mars Society, 240
Martian Chronicles, The (Bradbury), 238
Martians, 1–3, 41, 67, *74*, 74–75, *83*, 83–85, *84*, *86*, 86–87, *103*, 107, *110*, 110–13, *111*, *112*, 117, 131–33, 153–56, *154*, *156*, 158, 160, *181*, 185–88, *187*, *188*, 191, 197–98, *205*, 228, 230, 236
Marvin the Martian, 3
Massachusetts, *10*, 87, 97, *98*, 99, 117, 147, 155, 174, 177, 189, 202, *202*, 235
Massachusetts Hall (Harvard University), 9–10
Massachusetts Historical Society, 15–16
Massachusetts Institute of Technology (MIT), *14*, 14–15, 58, 59, 64, 122, 159, 177
mass hysteria, 183–84
Maunder, Edward Walter, 120–22, 125–29, 143, 151, 171, 207, 211, 219–21; *see also* "Small Boy Theory"
McCall, Mrs. William W. *See* Struthers, Irva
McGee, WJ, 152–53
McKinley, William, 85, 88, 97
Memoirs of Extraordinary Popular Delusions (Mackay), 183–84
Mercury, 27
Message from Mars, A (Broadway show), 111–13, *112*, 268
meteorites, 239
meteor showers, 27, 38
Meudon Observatory (France), 211–16, *212*, *215*, *220*
Mexico, 75
Mexico City, 75
Milan, Italy, 36–37, *37*, 54, 73, 77, 141, 217
Milky Way, 143
Min Yong Ik, 18, *18*
Model T, 192
moon, the, 3, 26–27, 38, 42–44, *43*, 54, 65, 98, 99, 123, 168
Morgan, J. P., 2, 146–47, 175, 192
Morse, Edward S., *14*, 14–15, 17, 35, 59, 117, 147–48, *149*, 151, 155–56, 227, 276
Morse code, 92, 95, 123, 176, 198
Mount Fuji, 56
Mount Hamilton, 32, *32*, 41, 76
Mount McKinley, 204–5, 218, 220
Mount Wilson Observatory, 203, 219, *220*, 222
Museum of Fine Arts (Boston), 58
My Favorite Martian (television sitcom), 3
mystiscope, 1

NASA, 3, 240
National Geographic Society, 163, 177, 218
Native Americans, 48, 95, 110, 145
natural selection, 12, 80, 84, 209
Nature, 155
neurasthenia, 77, 79, 91, 97, 225–26
Newfoundland, 118, *118*, 123
New Golden Hours, The, 108–9, *109*
New Jersey, 66, 189, 231
New Mexico, 157–58
New Nationalism, 224
New Orleans, Louisiana, 64, 111
Newton, Sir Isaac, 42
New York City, 1, 17, *17*, 33, 66, 86, 90, 91, 101, 105, 106, 112, 117, 124, 130, *130*, 140, 145–47, 157, 158, 166, 173, *175*, 177, 185, 192, 224, 232
New York Evening Journal, 86, 87, 91
New York *Evening Post,* 88
New York Herald, The, 21, 27–28, 34–35, 40, 41, 75, 106–7, 161, 197–98
New York Hippodrome, 1
New York Journal, 86, 94, 105, 107
New York *Press,* 36
New York *Sun,* 44, 66, 124, 149, 173, 201
New York Times, The, 2, 34, 107, 142–43, 149, 160–61, *161*, 170, 172, 173, 174, 184, 192, 198, 204, 208, 226, 227
New-York Tribune, 18, 193, 200
New York *World,* 2, 32–33, 35, 40–41, 86, 125
Nile River, 63
North Africa, 70, 73–75, 99
North Pole, 163, 204–5, *205*, 209–10, *210*, 217–18, 221

Oakland, California, 41
obsessive-compulsive disorder, 95
Oficina Alianza. *See* Alianza, Chile
O.K. Corral, 49
Olympic Games (St. Louis, 1904), 145
Opportunity rover, 239
opposition, astronomical, *29*, 29–31, 39, 47, 107, 123, 142, 144, 165, 191, 198, 204, 211, 233
Optique, L' (Flammarion), 119, 129
Our American Cousin (play), 89

Paine, Thomas, 184–85
Painted Desert, 148

painting, 224
Palazzo di Brera (Milan), 37, *37*, 223
Panama, 167–68, 173
Panama Canal, 163, 167, *167*
Panic of 1893, 48
Panic of 1907, 173, 177
paranormal, belief in the, 131–32
Paris, France, 23–25, 31, 54, 68, 69, *69*, 81, 119, *132*, 141, 142, 209–14, 216, 219, 225
Paris Observatory, 24, 69
Parker, Horace (fictional character), 111–13
Pasadena, California, 237
Pascal, Blaise, 190
Pathfinder rover, 239
Pau, France, 22–23
Peabody, Frank, 16
Pearson's Magazine, 82
Pears' Soap, 110, *110*
Peary, Josephine, 163–64, *164*, 166
Peary, Robert E., 163–64, *164*, 173, 177, 204–5, 210, *210*, 218
Peep at Mars, A (skit), 111
Pennsylvania, 31, 139
Percival, Sir (legendary character), 58
Perseverance rover, 239
Persians, 29
Peru, 31, 40, 168, 172
Philadelphia, Pennsylvania, 117, 139, 189, 206
Philadelphia Inquirer, The, 106
Philippines, 88, 153
Philosophical Basis of Evolution, The (Croll), 35
Phoenix, Arizona, 49
photography, 119–20, 143–44, 169
physics, 42, 131, 224–25
Pickering, Edward C., 45, 47, 184, 207–8, 282
Pickering, William H., 47–48, 51, 114, 198–99, *199*, 208, 259, 281
Picture of Dorian Gray, The (Wilde), 22
Pierre Guzman Prize, 23, 106–7, 191, 206, 232, 256
Pikes Peak, 95
Pinacoteca di Brera (Milan), 37
Planète Mars et Ses Conditions d'Habitabilité, La (Flammarion), 45
planets, 11, 24, 26–28, 61, 143, 153, 185, 224, 238

Plurality of Inhabited Worlds, The (Flammarion), 24, 26
poetry, 20, 23, 56, 57, 58, 97, 117, 139, 224, 225, 226
polar ice caps (Mars), 30, *30*, 31, 47, 52–53, 57, 123, 136, 137, 150, 158, 162, 164, 170, 241
Popular Science Monthly, The, 45
Portland, Maine, 190
Portland, Oregon, 228
Portland *Oregonian,* 66–67
power transmission, 92, 146
Primrose Hill (London), 84
Princess of Mars, A (Burroughs), 236–37
Proctor, Mary, *197*, 197–98
progress, belief in, 12, 109
Providence, Rhode Island, 190
Puerto Rico, 88
Pulitzer, Joseph, 32, 86, 155
Putnam, Elizabeth Lowell, 13, 35, 114
Putnam, William Lowell, 35, 45, 114–15, 123, 170

Quest and Achievement of the Holy Grail, The (mural), 58

racist attitudes, 20–21, 145, 152–53
radio, 2, 82, 92, 94–95, 123, 131, 176, 199–200, 233, 234
radioactivity, 224
radium, 224
ragtime, 110
religion, 12, 55, 65, 184–86, 190–91, 255
retrograde motion, 29
Richardson, H. H., 58
Richardson, Robert S., 236
"Rite of Spring, The" (Stravinsky), 224
Rochester, New York, 55, 143
Rochester *Democrat and Chronicle,* 55, 205–6
Rogers Building (MIT), 14, *14*, 59, 159, 255
Rolls, Charles S., 193–95
Rolls-Royce, 193
Roman Catholic Church, 185
romances (adventure novels), 62, 109
Romans, ancient, 29
Roosevelt, Alfred, 34–35
Roosevelt, Katharine Lowell, 34–35, 85, 116

Roosevelt, Theodore, 16, 34, 85, 88, 101, 120, 157, 163, 164, 167, 224
Rotch, Abbot Lawrence, 194–95, *196,* 201
Rough Riders, 88
Royal Hospital School, 121–22, *122,* 125–28; *see also* "Small Boy Theory"
Royal Naval College, 121
Royal Observatory (Greenwich, England), 120, 121, 143, 151, 171, 219
Russell, Ada Dwyer, 224

Sagan, Carl, 237, 238, 247
Sahara Desert, 63
St. Louis, Missouri, 131
St. Louis Republic, The, 145
St. Louis World's Fair (1904), *144,* 144–45, *152,* 152–53, 158
St. Moritz, Switzerland, 142
St. Patrick's Cathedral (New York), 86
Sands of Mars, The (Clarke), 238
San Francisco, California, 43, 78, 109, 164, 191
San Francisco Bay, 32
San Francisco *Examiner,* 41, 100, 106, *106,* 285
San Francisco Peaks (Arizona), 49, *49,* 228
San Juan Hill, 88
Saturday Review, The, 79
Saturn, 26, 45
Schiaparelli, Giovanni, *38,* 38–40, *39, 40,* 44–46, 51, 53, 54, 57, 63, 66, 70, 73, 77–78, 81, 85, 119, 125, 128, 131, 132, 134–35, 141, 158, 165, 189, 211, 213, 217, 219, 222–23, 238
science, 2, 12, 25, 42, 67, 72, 91, 190, 192, 208, 224–26, 230, 234
science fiction, 235–38; *see also specific works, e.g.: War of the Words, The* (Wells)
Science magazine, 206
Scientific American, 91
séances, 131–32, *132*
Selfridge, Thomas, 197
sensationalism, 32, 36, 44, 86, 89, 91, 155, 156, 209
Seoul, Korea, 18, 19, 138
Serviss, Garrett P., Jr., 145, 190
Serviss, Garrett P., Sr., 43–45, *44,* 54, 65, 67, 74, 87, 94–95, 107, 117, 122–23, 128, 142, 156, 190, 222, 225, 229, 234–35

"Signal from Mars, A" (two-step), 111, *111*
Signal Hill (Newfoundland), 118, *118*
Silver Pearl (alleged Martian woman), 131
Singer Tower (New York), 192
Sion College, 218–19, *219*
Skygack, Mr. (cartoon character), 186–87, *187*, *205*, 224
"Small Boy Theory," *125*, 125–29, *126*, 143, 151, 207, 211, 219, 271
smartphones, 146
Smith, Hélène (alias), 132–33, 135, 140
Socialist Party, 55
So Koang Pom, *18*
solar eclipses, 98–99, 165
solar system, 3, 11, 26, 27, 29, 45, 57, 61, 156, 186
Somerset, Lady Henry, 184
South Africa, 82
South Dakota, 35
South Pole, Martian, 47, 52–53, 137
space opera, 236
Spain, 85–88, 94, 106
Spanish–American War, 87–88, 94, 106, 147
Spencer, Herbert, 12
Spirit rover, 239
Spokane, Washington, 228
SS *Colon*, 173
SS *Panama*, 166
Stanford University, 228
Stevens, Leo, *200*, 200–201, 286
Stevenson, Robert Louis, 62
stock market, 48, 164–65, 192
Stravinsky, Igor, 224
strikes, 22, 31, 55
Struthers, Irva, 139, 231
strychnine (as a treatment for neurasthenia), 77
Sullivan, Anne, 145
sun, the, 11, 26, 27, 29, 30, 36, 47, 55, 98–99, 121, 155, 165, 198, 199
sunspots, 26, 27, 121
Süring, Reinhard, 200, 201
Surrey, England, 79, 81, 195
survival of the fittest, 57
Switzerland, 132, 142, 194

Taft, William Howard, 199
Tarzan novels, 236
telepathy, 131, 233
telephones, 95, 109
telescopes (in general), 30–32, 40, 45–48, 53, 70, 120, 127, *149*, 164–65, *169*, 216, *220*; see also specific observatories, e.g.: Lick Observatory
Terby, François, 39, 40, 45
terrorists and terrorism, 22, 31
Tesla, Nikola, *90*, 90–97, 101, 105–10, *106*, 113–14, *124*, 124–25, 145–47, 160, 167, 174–76, 191, 199, 223, 232, 265–67
"Things that Live on Mars, The" (Wells), 187–88, *188*
Times, The (London), 40
Todd, David, 97–99, *98*, 165–69, *169*, 171–74, 178, 179, 191, 199–203, *200*, *202*, 208–9, 222, 232–33, 277–78, 282, 286, 287
Todd, Mabel Loomis, 97–100, *98*, 166–73, *169*, 178, 179, *189*, 189–90, 201–3, *202*, 224, 232–33, 276–77, 279, 286
Tokyo, Japan, 20, 35, 45
"To Mars with Tesla" (magazine serial), 108–9, *109*
torpedo boats, 93–94, 175, 265–66
Total Eclipses of the Sun (M. Todd), 98
Treasure Island (Stevenson), 62
Tremont Theatre (Boston), 35, 43, 46, 65, 68
triangle of lights (on Mars), 33, *33*, 35–36, 40, 85, 106
Trinity Church (Boston), 58
Tripoli, Libya, 99, 165
Tucson, Arizona, 49, 115
tulip mania (Holland), 183
Turner, Herbert Hall, 72–73, 151, 153
Twain, Mark, 86, 90
Tyndall, John, 42

unconscious mind, 133
United States, 2, 18, 22, 31, 55, 85, 88, 109–10, 134, 163
University of Arizona, 49, 115, 139, 170
University of California, 228
University of Copenhagen, 218
University of Geneva, 132
University of Oxford, 72–73, 151, 153
University of Wisconsin, 206

Urania (Greek muse), 69–70
Uranus, 26
urbanization, 22
U.S. Army, 48, 197
U.S. Civil War, 14, 133
U.S. Congress, 157, 218
U.S. Federal Reserve System, 173
U.S. Geological Survey, 114
USS *Maine*, 87, 106
U.S. Supreme Court, 184
Ute tribe, 95

Valparaíso, Chile, 190
Van Anda, Carr, 143, 273
vaudeville, 111
Venus, 26, 27, *76*, 76–77, 113, 114, 127, 146, 263, 267
Verne, Jules, 156, 233
Victoria, Queen, 58, 86, 101
Victorian England, 12, 13, 67, 223
Viking spacecraft, 239
Vin Palmette (patent medicine), 110

Waldorf-Astoria Hotel (New York), 90, 105, 145–47, 173, 175, *175*, 232
Wallace, Alfred Russel, 209
Wall Street, 48, 165, 192
Wall Street Journal, The, 107, 173, 177–78, 184
Walston, Ray, 3
war, 72, 85, 159, 226, 285; *see also specific wars, e.g.:* Spanish–American War
Warden, James, 124
Wardenclyffe, Long Island, 124–25, 146, 147, 175, 232, 270
Warner, Charles Dudley, 68
War of the Worlds, The (Wells), 82–87, *83, 84*, 124, 187, 195, 233–35, 287
Washington, D.C., 48, 55, 197, 199
Washington *Evening Star*, 162
water, 27, 44, 53, 60, 61, 63, 71, 73, 135–38, 148, 158, 160, 209

Weiss, Sara, 131
Welles, Orson, 82, 234
Wells, Frank, 80–81
Wells, Fred, 82
Wells, Herbert George "H. G.," 79–87, *80, 83*, 155–57, 187–88, 226, 233, 236, 287; *see also War of the Worlds, The*
Westchester, New York, 34
Westinghouse, George, 92
West Point, 19
West Virginia, 35
Wharton, Edith, 20
White House, 19, 157
Whiting, Lilian, 64–66, *65*, 68, 81, 98, 100, 114, 160, 163, 164, 198, 202, 233, 261, 275, 287
Whitman, Walt, 30
Wilde, Oscar, 22
wireless (wireless technology), 4, 92–96, 107–8, 118, 123–24, *124*, 131, 146, 175–76, 187, 199, 221; *see also* radio
Woking, England, 79, 81–83, 195
women's clubs, 188–90
Woolf, Virginia, 223
World's Fair. *See* Chicago World's Fair (1893); St. Louis World's Fair (1904)
World War I, 3, 226–27, 233
World War II, 233
Wright, Orville, 192–93, 196–97, 199, 202
Wright, Wilbur, 192–93, *193*, 196–97, 199, 202

X-rays, 90, 131

Yale University, 179, 184
Yankee Circus on Mars, A (stage show), 1
yellow journalism (yellow press), 86–91, 94, 106–8, 142, 170, 174, 208, 224
Yerkes Observatory, 206–8, *207*, 216, 217
York, Maine, 139

ABOUT THE AUTHOR

DAVID BARON is an award-winning journalist, broadcaster, and author. Formerly a science correspondent for NPR and science editor for the public radio program *The World*, he has also written for *The New York Times*, *The Washington Post*, *The Wall Street Journal*, the *Los Angeles Times*, *Scientific American*, and other publications. While conducting research for *The Martians*, he served as the Baruch S. Blumberg NASA/Library of Congress Chair in Astrobiology, Exploration, and Scientific Innovation. An affiliate of the University of Colorado's Center for Environmental Journalism, he lives in Boulder.